GADGET NATION

A Journey Through the Eccentric World of Invention

STEVE GREENBERG

"The Innovation Insider"

STERLING

New York / London

www.sterlingpublishing.com

Every effort has been made to provide in these pages straightforward and accurate information regarding new and existing gadgets and products. The author and publisher cannot be held liable for any misuse or damages stemming from the content of this book.

STERLING and the distinctive Sterling logo are registered trademarks of Sterling Publishing Co., Inc.

Library of Congress Cataloging-in-Publication Data
Greenberg, Steve, 1960–
   Gadget nation : a journey through the eccentric
world of invention / Steve Greenberg.
      p. cm.
   Includes index.
   ISBN 978-1-4027-3686-5
   1. Inventions. I. Title.

T212.G565 2007
609—dc22
                        2007027672

10 9 8 7 6 5 4 3 2 1

Published by Sterling Publishing Co., Inc.
387 Park Avenue South, New York, NY 10016
© 2008, 2010 by Steve Greenberg
Distributed in Canada by Sterling Publishing
%o Canadian Manda Group, 165 Dufferin Street
Toronto, Ontario, Canada M6K 3H6
Distributed in the United Kingdom by GMC Distribution Services
Castle Place, 166 High Street, Lewes, East Sussex, England BN7 1XU
Distributed in Australia by Capricorn Link (Australia) Pty. Ltd.
P.O. Box 704, Windsor, NSW 2756, Australia

Book design and layout: *tabula rasa* graphic design
Art direction: Chrissy Kwasnik

Sterling ISBN 978-1-4027-3686-5 (hardcover)
         ISBN 978-1-4027-7799-8 (paperback)

For information about custom editions, special sales, premium and corporate purchases, please contact Sterling Special Sales Department at 800-805-5489 or specialsales@sterlingpublishing.com.

To my Dad, Lazarus,
who taught me to
respect ingenuity—
and to my Mom, Dorothy,
who taught me the joy
of listening to and learning
from everyone's life stories.

CONTENTS

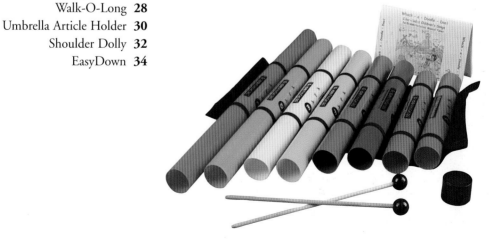

INTRODUCTION

We live in a world where medical miracles and high-tech marvels are almost commonplace. New cure-all prescription drugs fill our pharmacies daily, and last year's electronic products are this year's dinosaurs; as smaller, stronger, and smarter versions are launched each week. Most of us look at these innovations as something we could never invent. We're lucky if we even can figure out how to use them.

Every year, however, a number of inventions show up in the marketplace that make us say, "Why didn't I think of that?" I'm sure any one of us could have invented the paper clip. We just didn't do it.

Now, before I go any further, I want you to know that I'm an inventor groupie. I have nothing but respect and admiration for all inventors. I think it's part of my DNA. My Dad, Lazarus Greenberg, and his brother, Marvin, were always coming up with clever gadget solutions.

When my brothers and I couldn't reach the doorknob on the back door, my Dad added a lever and chain to the knob. We pulled on the chain and the door would open. Simple, but it worked. My Dad and Uncle also came up with springs for eyeglasses long before they showed up on fancy designer frames. Here are a few actual sketches showing their improvised solutions to the prob-lems of washing-machine lint, vent pipes, and clogged gutters.

Sadly, I wasn't raised on royalties from inventions. I don't even know if they ever filed for a patent. But despite their lack of monetary success in that field, or maybe because of it, I became hooked on invention, ingenuity, and innovative products. Like I said, I guess it's just in the DNA.

These days, I earn part of my living as the "Innovation Insider." I travel around the country showcasing some of America's smartest new products. In my

mind, there's nothing more American than that entrepreneurial spirit that all inventors share. In this country, if you hit the right idea, your life and the lives of your children, grandchildren, and even great grandchildren will never be the same.

Approximately 60,000 patents are filed each year by private Americans. That's a lot of people grabbing at that brass ring. When I talk to inventors, many working out of their garages, I'm always impressed with their passion and persistence. They truly believe they have the next must-have product. Friends and family may tell them they're nuts, but they push forward. And almost all of the inventors I have spoken to have said they want their products to be a success, but not only because of the money. They genuinely want to change the world. There's an indescrib-able joy in seeing your invention sitting on a store shelf. It's like having a bit of

Some of Dad's sketches

PREVENT BIRDS AND LEAVES FROM CLOGGING EXHAUST

PREVENTS CLOGGED DRAINS

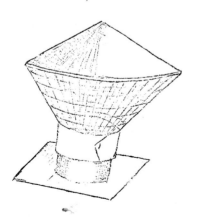

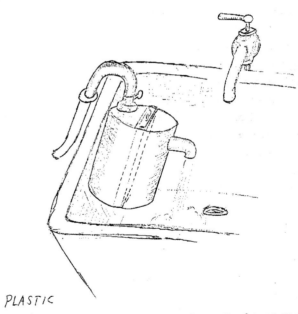

VENT PIPE COVER

PLASTIC WASHING MACHINE LINT CATCHER
WITH DISPOSABLE FILTER

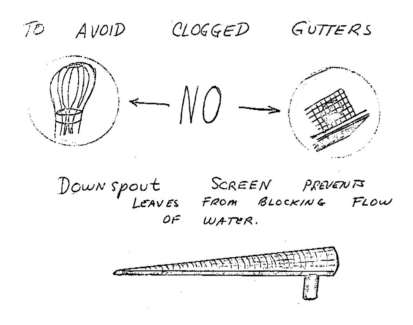

TO AVOID CLOGGED GUTTERS

NO

DOWN spout
LEAVES
OF

SCREEN PREVENTS
FROM BLOCKING FLOW
WATER.

America's less conventional gadgets and talked to the inventors behind them. You'll meet men and women who had an inspiration and are now turning that idea into something we can all buy. You'll find out why some of your neighbors have put their passion, their energy and often a great deal of their own money into these unusual gadgets.

I could never tell, by just hearing about a product, if that product was going to be a financial success or not. It was only after talking with each of these inventors that I discovered which products are actually making it and which ones are not. Some products, which I might have categorized as lame, are making money. Other products, which I thought were very clever, have tanked. That was part of the fun of writing this book. The backstories are all surprises. Talking to garage inventors is a wonderful way to tap into modern-day Americana. They are the grassroots of

consumer immortality. All inventors believe their problem-solving products will make the world a better place.

That said, not all new products are destined to make the world a better place. Some inventions are just plain . . .

bizarre. Others are clever in some strange way. And that's what this book is all about. It's a collection of some of the off-beat products out there. And when I say "out there," I mean "out there." I've rounded up some of this country's ingenuity and entrepreneurial spirit. Whether the product is a great idea or a dud, you'll find that spirit is strong.

Some of these inventions, however, are examples of innovation gone terribly wrong—good people/bad gadgets. After reading about some of these painful experiences, you may find yourself breathing a sigh of relief, saying "Boy, am I glad I didn't think of that!"

But it's in these less-than-great gadgets, and the stories behind their creation, that you'll find ample illustration of the indomitable passion that feeds the spirit of invention. At the same time, they provide a funhouse mirror of some uniquely American attributes: our obsession with convenience, the fascination with gimmickry, and the eternal dream of getting rich quickly. Have the Chia Pet, The Clapper, Ab Twister, or The Miracle Mop made the world a better place? They have certainly made their

inventors a great deal of money and become part of our consumer pop culture. That's the dream of the garage inventors profiled in this book: Leave their mark in the marketplace and make a fortune doing it.

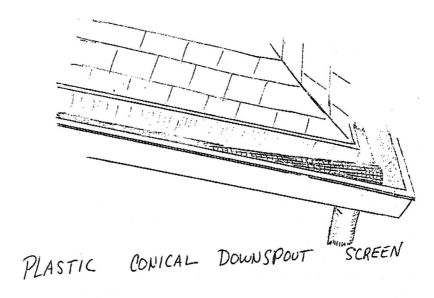

PLASTIC CONICAL DOWNSPOUT SCREEN

Mo-Bowl Mobile Pet Water Bowl
Help Alerter
Cast Skate
Hold-a-Phone
Take It Pocket
Drink Deputy / Travel Mate / Drink Mate
Diaperbridge
Re-Pillable Card
Mother's Third Arm
Air Guardian
OrthoTote
Take-Out-Time-Out Mat
Roadside Message Board
Walk-O-Long
Umbrella Article Holder
Shoulder Dolly
EasyDown

1

On the Go

We want it fast, we want it now, and we want it on the go. Americans love to be on the move, but hate the hassles that go along with it. It's that hunger for convenience that inventors feed off. They know we will crack open our wallets to make our lives just a little easier, a little more hassle-free. With all its complications, travel is the perfect feeding ground for inventors. There are so many bumps along the road, so many annoying and frustrating obstacles associated with driving and flying—airports, luggage, kids, pets, drinking, diapering—I'm getting cranky just thinking about it all. Inventors focus on those hassles and try to find solutions that we might all be willing to buy into.

One of the inventors in this category is a mom who needed a way to punish her kids when the family was away from home. At home, she could send them to the dreaded "time-out chair." On the road, that chair wouldn't fit in the car—and the kids knew it. Lisa Bogart Carvajal came up with an idea that gives her the upper hand even when she's nowhere near a time-out chair.

From giving water to the family dog while driving down the highway to carrying medications in your wallet, these inventors have found a way to get you from Point B to Point C without the usual pain in the A.

"...everything either looked good but didn't work, or worked but was huge"

Mo-Bowl™
Mobile Pet Water Bowl

Bowl Me Over

Inspiration comes in many shapes and sizes. For Rich Skowronski, his inspiration had four paws and long hair. "My dog, Bonnie, demanded it."

Rich is talking about his invention, the Mo-Bowl Mobile Pet Water Bowl. You see, Bonnie, Rich's golden retriever, must have water while going on any car trip. "When I moved to New Hampshire, I would often take Bonnie in the car with me for several hours, and I'd bring along water for her, but it would spill." The final straw occurred when Rich took Bonnie on vacation and the water bowl spilled right onto the dog bed, leaving Bonnie with a soaking wet bed. That's when he came to the conclusion there had to be a better way, or at least a drier way.

Rich is an engineer with twenty-five years of experience. "I looked to see what was on the market," he says, "and everything either looked good but didn't work, or worked but was huge"—way too big for an ordinary car.

"Dogs have a spectacular need for water, and not just in the heat. Even in moderate or cool weather, dogs need a lot of water in the car. I felt I could fill a need that a lot of people didn't even realize their dogs have: water is so important for both their safety and comfort."

Rich already had "20-something" patents to his credit, including the flexible deck on treadmills and the cord that attaches telephones to the back of airplane seats. But it was the idea for a mobile pet

water bowl that led Rich to give over the management of his engineering business to his wife and devote himself full-time to developing a no-spill pet water bowl.

Inspired by travel coffee mugs, which are well-designed to prevent spills, Rich developed a design to fit in an ordinary car cup holder and assembled his own prototypes, which he and his wife test drove in their Jeep Wrangler along their half-mile-long bumpy driveway. "But nothing worked well. As an engineer, I wanted a product that would work perfectly."

Using his engineering skills, Rich was able to build prototypes quickly, easily, and inexpensively. After setting up a wooden platform test bed to test his designs, Rich mounted a video camera at very close range to record the tests, and then watched the video frame by frame to find the key to a no-spill design.

After a few final changes, Rich solved the mystery. "I'm amazed at how well it works now." And most important, "Bonnie loves it."

The Mo-Bowl went into production in September 2005 and is now available on the Internet and at pet stores. Rich's next steps include getting the word about the Mo-Bowl directly to consumers via media coverage. In one successful appearance, Rich and Mo-Bowl were featured on a Home & Garden television show. He got the lead for this show through a connection he'd made at a local inventors' club, a networking opportunity he advises other inventors to pursue.

As for additional advice for would-be inventors, Rich offers these words of caution: "If you have an idea and you want to turn it into an invention, it may take a couple of years of your life full time—that is, if you're lucky. It will require money, so be careful not to waste $100,000 in the process."

Bonnie does a "spill-free" demonstration.

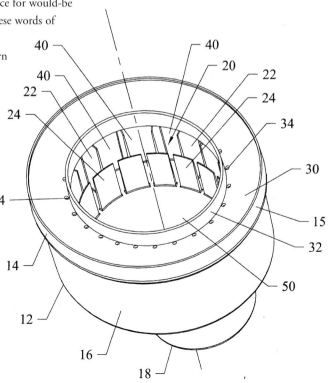

Help Alerter™

He Saw the Light . . . and It Was Flashing HELP

He looked to the heavens for help and found his answer in giving "help" to others.

When Dave Meester prayed for help, he got the Help Alerter. Dave said something from above gave him the idea for a license plate holder that quickly converts into a flashing Help sign.

Dave Meester had been downsized out of a job six months before, and his spirits needed a lift. He looked to the heavens for help and found his answer in giving help to others.

Dave thought of the time his wife, Barbara, had to pull her car to the side of the road and wait nearly three hours before anyone stopped to help her. Add to that the frequent carjacking stories in the news, and Dave realized the need for drivers to have some way to safely alert others that they need help.

Dave doesn't think of himself as an inventor. He credits the idea of the Help Alerter to divine inspiration and went straight to his workshop. Weeks later, he came out with a license plate holder that quickly converted into a flashing Help sign.

He made the first prototype case out of plywood, borrowed the pivot arms from pieces of an old fishing tackle box, and used dome lightbulbs for lighting. Since then, he's refined the design and spent serious money to have good prototypes handmade by a prototype builder. He plans on offering two models—one that is manually activated by getting out of the car and pulling the license plate down, and a more expensive model that is activated inside the car with a button on the floorboard.

"Say you're being carjacked. If you have a cell phone, could you use it? If you have OnStar, could you use it? Probably not without jeopardizing your situation. A silent button on the floorboard that activates the Help Alerter might be the only thing that would help

to report a stranded motorist, even if they didn't stop.

Dave explains, "A lot of people, especially women, are afraid to stop to help. But when they see that flashing request for help, they *will* make a phone call."

Dave's not yet at the point of having the Help Alerter manufactured. That's the next step. I'd guess I have $60,000 to $75,000 invested. A good chunk of that went to three marketing groups, who were supposed to make it a reality and get it to market. None of them produced a good result. That's why Richard, a retired friend, and I are doing it ourselves. Richard wants to get involved because he feels it will be a hot item. I think they'll go like hotcakes."

Of 250 people Dave surveyed, 80 percent said they would buy the Help Alerter when it becomes available. He's got believers in the local sheriff's department, too. The York County, South Carolina, sheriff's department wants to do a TV commercial to endorse it. Dave knows an official endorsement can go a long way in alerting the public and advancing this venture.

you in this situation. This could be the only chance you have. People would see the flashing Help sign and call the police."

Dave spent a year of weekends and evenings doing his own market research. "I would find a stretch of road where I'd pretend my car had broken down. I'd pull over to the side of the road, use my flashers, sit inside the vehicle—and wait. The average time I'd sit there before anyone would stop was twenty to thirty minutes. Then I'd do the same thing using the Help Alerter. My wait time was between eight and ten minutes—a drastic reduction in time! Plus, a police officer would stop more often when I used the Help Alerter than when I didn't."

He discovered that the higher police officer involvement was a result of people calling 911 on their cell phones

Dave didn't think he could get a patent—he says he's no genius. He was sure someone else had already come up with this simple idea. But the patent was his for the taking. Perhaps there was a bit of divine intervention for this invention. After all, Dave did get a sign.

A sign from above

"...I had no relationship with crutches. I could use them, but I just couldn't accept them...."

Cast Skate™

Cast Away

When Bob Bentivegna was forced to take an early retirement from the Jersey City Fire Department, he didn't go without a fight. Bob was a 51-year-old local boxing champion who'd won in the first round against both the New York City Fire Department and Police Department, and took three rounds that ended in a decision against the Newark Fire Department. You don't mess with Bob.

Bob may be tough, but his feet were not. Twenty-three years of getting in and out of fireman boots with stiff bunker pants on took their toll. His feet were deteriorating. The doctors took X-rays, MRIs, and bone scans, found a stress fracture in his ankle, and told him he couldn't go back to work. No more jumping off the rig for this guy. He was also sentenced to wear a rigid cast for six weeks.

For most people, this would be a major inconvenience; for Bob it was intolerable. "The problem was that I had no relationship with crutches. I could use them, but I just couldn't accept them. . . .

I cut them up and threw them in the garbage, thinking, 'I can't do this; there's got to be a better way.'"

It was a warm Indian summer night and, already restless, Bob couldn't sleep. Out of desperation, he crawled backwards down the stairs to his basement in the middle of the night to make something to help him get around on the cast. Looking around the basement for parts, he took apart the

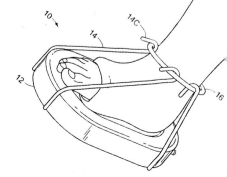

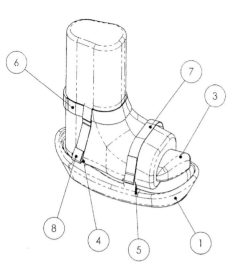

headrest from a sit-up bar and removed its oblong cushion. He then cut two bungee cords in half and screwed them to the plastic bottom of the headrest. Next was determining how to anchor the cords.

Looking around again, he saw the rigid brown drive belt from a vacuum cleaner and slipped it over the toe of his cast. He rolled the drive belt up over his ankle and attached it to the bungee cords. It worked. The Cast Skate was Bob's ticket to freedom for the next six weeks. He could pivot off his bad foot and lean on his good leg. The Cast Skate was easy to put on and take off. When Bob went to bed, he just kicked off the four cords, keeping the cushion on the bottom of his cast.

Six weeks later, when Bob got the cast off, he showed his Cast Skate to his orthopedic surgeon. The doctor examined Bob's invention and liked it. A patent search revealed that no one else had a patent on this idea. So that meant Bob's feat wouldn't step on anyone's toes. In fact, its construction was so unique that he was able to get a utility patent, not just a design patent, and his patent was granted on the first application. In the world of patents, that's a big deal.

Next, Bob's invention needed some attention, which is difficult for a single inventor to get. So he went to the Yankee Invention Expo in Connecticut. A few months later, he received a couple of letters from interested manufacturers. "My wife and I were doing backflips. But we still have to sit back and be patient. Nobody hurries in this business. They just don't. It took a year and a half to get my patent and that's like overnight. No one has a sense of urgency about my product, except for me."

Today, Bob gets promo material in the mail and always responds to licensing inquiries. But he knows to be suspicious of anyone who asks for money up front. "Some of these people want $6,000, plus 20 percent of the proceeds, and have you sign over your patent for the full twenty years. I'm not comfortable with that."

Bob has already spent about $6,000 on his invention, most of that on his legal fees. "It's not wise for someone in my position to dish out thousands of dollars. The product needs to speak for itself."

Bob has read an armload of books on the invention and patent process. He doesn't skate around the fact that the Cast Skate is still in its early stages. "One of my books says, 'Even an overnight success takes a long time.' If you're an independent inventor, you have a long road ahead of you."

And Bob and his Cast Skate are heading down that road one step at a time.

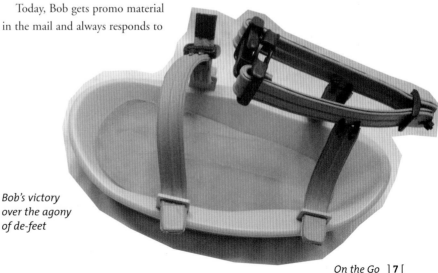

Bob's victory over the agony of de-feet

STAT BAR

PATENT: pending

PRODUCT PRICE: $19.95

STATE: Massachusetts

INVENTOR'S AGE: 51

INVENTOR'S PROFESSION: sales

MONEY SPENT: $100,000+

MONEY MADE: N/A

WEB ADDRESS: hold-a-phone.com

In just three years, Jim has gone through all of his savings to create his Hold-a-Phone.

Hold-a-Phone™

Excuse Me, Your Wrist Is Ringing

After missing calls and a broken dropped phone, Jim McGrath wondered. Is there any possible way to keep a cell phone handy and still have your hands free?

That's the question 51-year-old Jim McGrath had been asking himself. Being in sales, Jim needed his cell phone to be accessible—no holds barred. So he studied several types of phone holders on the market. Some models would hold a phone, but had to be taken out of the holder to be used. He thought about bicep armbands, but the phone was too high to see the caller ID or to talk into it in that position.

So Jim decided on the wrist position and focused on the two main styles of cell phones: bar and flip phones. The idea was to be able to talk without taking the phone out of its holder and yet be able to see the caller ID. Choice of material narrowed quickly. He chose neoprene, which is soft and wear-able, lends shock absorbency to the phone, and stretches,

allowing one design to accommodate several sizes and brands of cell phones. And its pliability allows the user to dial through it.

In just three years, Jim has gone through all of his savings to create his Hold-a-Phone. "Luckily, I only had to make two prototypes. The first company I used was in Taiwan. They almost got it right the first time. I only made a few changes and it was done. I found Korea's manufacturers more expensive than China's. Now I get them made in China and save one dollar per holder. But I

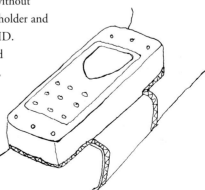

Hands-free phoning

have to stay on top of the Chinese manufacturer—the quality control can get sloppy. I'm up late every night instant messaging to China, and working on new inventions."

Thankfully, after investing $100,000, he doesn't have to spend money on it now. He's gotten a lot of free publicity. *Runner's World* magazine ran two sentences about it, and even without a photograph, he sold more than 200 that month. He's just starting to buy advertising.

"My customer base is 80 percent women. That percentage was higher, around 95 percent, but now men are coming around. Women buy them for safety and security reasons. When they exercise, they want to bring their phones and need a place to put them. Men are buying them so they can use their phone during a race; that way they can call their wives and relay where to pick them up."

The Hold-a-Phone has been particularly helpful to the physically challenged.

As one buyer said, "My husband is confined to a power wheelchair and only has limited use of his right hand (his left hand and arm are immobile). We strap the Hold-a-Phone to his left wrist above his wristwatch. This enables him to make and receive calls using only his right hand and a Bluetooth wireless earpiece. . . . We love it!"

Jim's marketing strategy is to contact everybody—there's no holding him back. The Hold-a-Phone is sold on his website and in a few catalogs and stores. It's been a challenge to get it into retail stores. Sporting goods stores say it's a cell phone item and cell phone stores say it's a sports item. Without his aggressive marketing, he'd be left holding the bag—or, phone.

To take hold of the market, Jim bought all of the Internet domains related to the phrase "hold a phone," hyphens or not, and he worked his website to come up number one on Google searches for "arm phone holder" and "wrist phone holder." Once people have one Hold-a-Phone, they buy more for their friends and running mates. Most just don't know it's out there.

By moving the cell phone from the belt to the wrist, Jim McGrath hopes to change the way we communicate. With Hold-a-Phone, this inventor has put a lot on the line, but it's a "wrist" he's willing to take.

Take It Pocket™

It's a Sign

STAT BAR

PATENT: pending

PRODUCT PRICE: $8.49

STATE: Washington

INVENTOR'S AGE: 35

INVENTOR'S PROFESSION: product designer

MONEY SPENT: $25,000

MONEY MADE: $0

WEB ADDRESS: takeitpocket.com

"Used car sales is the obvious use for Take It Pocket, but activism is not as obvious. That's why I'm doing it."

Jack Huang and his wife, Flora, had just purchased a new car and were trying to sell their old car. "Using plastic 'For Sale' signs didn't interest me because they blocked my view while driving, and they were simply unattractive." So Jack designed a flier to post on bulletin boards and put up on telephone poles.

Flora, a real estate agent, had several of those bulky acrylic Take One literature holders around the house. "I think somewhere between seeing those holders at home and having just shopped for a car and noticed the metal window key boxes auto dealers use, the idea for a Take It Pocket entered my head."

Jack cut up some clear vinyl folders into a bunch of vinyl sheets. He then taped the vinyl sheets together and made four prototypes, mostly to try different materials. Jack's Take It Pocket is a clear plastic sleeve that attaches to a car window. It has two pockets, one that can hold a message and, if you lift the message panel, a second that can hold 8½-by-11 fliers.

Since the car sold right about the same time Jack made the prototypes, he didn't get to use his product. "The idea went into my drawer of ideas. And for the next three years I didn't do much with it."

In late 2005, after leaving his job at Microsoft to start a new venture, Jack was digging through the drawer and found his old Take It Pocket. "I started gathering data about the number of

This invention can be used to sell a car—and also much more . . .

Jack regards cancer awareness as a good fit for two reasons: there's breast cancer in his family and, as it happens, Lance Armstrong's LiveStrong cancer awareness wristbands are yellow. Jack's last name Huang means yellow. For Jack, that's a sign.

"It's a new way to talk about issues. I love the idea of making a difference . . . to leave something behind that *means* something is powerful for me."

If you don't think a few sheets of clear plastic can change the world, Jack Huang isn't worried. He has confidence in his Take It Pocket. If you don't agree, it's an opinion that you can Take It or Leave It.

registered cars and the various ways people use fliers, and came to the conclusion that this could become a new way to communicate. Just as Post-it Notes changed the way we communicate, so could my Take It Pockets."

Jack says he saw his product as more than a silly way to help sell used cars. People could use their parked cars to sell a point of view. For example, his Take It Pocket could be used to elect candidates, raise money for charities, or make people aware of issues.

People he showed his invention to would say, "Are you sure this doesn't already exist?" But Jack checked—and it didn't. So he filed for a provisional patent.

Jack then started to look for local plastic manufacturing companies that could produce his Take It Pocket. Next, he tested many different car sleeve plastics on his back deck to see which ones could tolerate sun, wind, and rain.

Now, Jack has already manufactured about 5,000 Take It Pockets and is about to launch this product. Instead of pursuing the used car sales route, he's teaming with *KnowCancer.org* to use his Take It Pockets to raise cancer awareness. "Used car sales is the obvious use for Take It Pocket, but activism is not as obvious. That's why I'm doing it."

Drink Deputy™/Travel Mate™/Drink Mate™

"Look, Mom, No Hands"

"I would love to say we're kicking butt, but we're not."

Denny Kays heard the concerns of a grandma who watched her grandson repeatedly throw his "sippy cup" while her daughter was driving—and then watched with dismay as her daughter repeatedly turned around to pick it up! A crash just waiting to happen . . . but not with our hero Denny around.

He thought about the problem and came up with Drink Deputy, a harness to hold a bottle or sippy cup. If baby tossed the bottle, it would stay tethered to the car seat, stroller, high chair, grocery cart, or whatever. The bottle wouldn't hit the floor and wouldn't get lost.

This 59-year-old had worked in sales all his life, first selling doors and more recently baby photographs. It might have been that baby connection, but whatever it was, he took the idea and ran with it.

It took at least fifty prototypes to find the right design and the right elastic that would fit most bottles and sippy cups. The memory elastic he found has a ten-year guarantee and won't let even the largest sippy cup slip out. That's a good memory.

After having Drink Deputy evaluated by the Juvenile Products Manufacturers Association (JPMA) and Consumer Product Safety Commission, Denny was ready to put this Deputy on duty.

He initially showed it at a baby fair in Chicago. JPMA and ABC Kids Expo and other baby-kid shows are how he got his business rolling. Buyers liked it, and sales started

Bottle bye-bye no more

to crawl in. Today, an exclusive distributor sells it in twenty-six states and Canada. Specifically, it's sold in baby boutiques, gift stores, hospital gift shops, by Web businesses, and in mom-and-pop stores.

While the Drink Deputy has benefited from free publicity in newspapers, ads in magazines, and TV time, sales have been just so-so. "I would love to say we're kicking butt, but we're not. We need to have a much bigger advertising budget than we have. That's the struggle. The second is the difficulty of getting it into regional stores—so many stores don't want to deal with a one-item vendor."

The idea of developing a Drink Deputy for adults hit him after giving samples of the Drink Deputy away at a trade show. Later, at the airport, Denny saw at least fifteen people from the trade show with a water bottle attached to their suitcase—using his Drink Deputy!

As soon as Denny got home, he redesigned his Drink Deputy so that it would attach to a belt, suitcase, or purse. The adult Drink Deputy would handle the popular 20-ounce water bottles and accommodate other sizes, too, with a top that fit over the cap, keeping the bottle upright. Denny decided to call this Drink Deputy for adults Travel Mate.

The idea to put various company names and logos on these drink holders came from a cheerleading coach chaperoning girls at a competition. Seeing his Travel Mate, she decided that every girl on the team had to have one "because the publicity would be great for school spirit and it would be a good fundraiser."

Denny began selling the Travel Mate for major fundraisers. He uses a pink ribbon design for breast cancer walks and runs (his wife Penny is a ten-year breast cancer survivor). Because people carry a water bottle more than they wear a certain T-shirt, the cause's name and logo get more publicity than standard giveaways provide.

Denny's Travel Mate now has the U.S. government's attention. The Department of Agriculture Forest Service is interested in it for firefighters—it's hard to keep them hydrated. VA Health Care sees how great it could be for people in wheelchairs. And the Army and Marine Corps recruiters think Travel Mates would be a better giveaway than their traditional coffee mug and T-shirt.

Denny sent a sample of Travel Mate to 300 random readers of *North American Fisherman* magazine and received an 81 percent approval rating from fishermen. So Denny responded by designing Drink Mate with the outdoorsman in mind. It securely attaches to a chair, tree stand, belt, or backpack. No outdoorsman should ever go thirsty again.

Denny knows the secret to his success is more publicity and he wishes he had the money to buy more ad time. He wanted to go on QVC television, but was told they don't sell baby products.

Denny is still working it and, if he has his way, the Drink Deputy and Travel Mate will find their way into every American home.

Drink Deputy, we copy that, 10-4.

Diaperbridge®

Is the Diaperbridge a Bridge Too Far?

STAT BAR

PATENT: US #6918147

PRODUCT PRICE: information not provided

STATE: New Jersey (but moving to Maryland)

INVENTOR'S AGE: 39

INVENTOR'S PROFESSION: pharmaceutical corporate lawyer

MONEY SPENT: $200,000

MONEY MADE: "not much"

WEB SITE: diaperbridge.com

"Can't someone invent something that can cover up the sink hole?"

Garrett Stackman could be nominated as Dad of the Year. Not only does he change diapers, this 39-year-old likes it. "I like to be involved. There's an intimacy in doing everything for your baby, even if it smells bad." And it's this strange love of diaper changing that gave birth to Garrett's portable changing station.

The Diaperbridge story started when Garrett and his wife, Lisa, were visiting in Maryland and went to a fancy restaurant for dinner. While there, Lisa took Alex, their three-month-old son, to the bathroom to change his diaper. Finding no place to change him and realizing it was too cold to go out to the car, she changed Alex on the floor. Yuk. Lisa voiced her anger at the situation when she returned to the dining table. "There's a perfectly nice vanity in there. Can't someone invent something that can cover up the sink hole?"

"I'm pretty handy," says Garrett, "so as soon as we returned home, I gathered materials." He laminated pressboard and a stop hinge so that when the Diaperbridge was unfolded, it made a sturdy platform. Then he attached harnesses for the baby and was ready to test it out. "When we put it over a sink, we realized we were sunk. If the baby grabs the faucets, he or she can get wet—or worse, scalded. And while a baby would be securely affixed to the platform, the platform itself could fall off the counter. The last thing we wanted was a baby getting hurt." So Garrett made a flip-up panel that covered the faucet, and added additional straps. "I used a 4-point harness to hold the baby in place. Most harness setups are 5-point, but you need the crotch-point open to change a diaper." Yup, that's a critical point.

Once they had a viable prototype, Garrett and Lisa searched for plastic manufacturers. Through the Internet, they hooked up with a top plastic guy near their home in Morristown, New Jersey. He designed a mold that can make half a million pieces. They started the patent process and the patent was issued on his wife's birthday. "I'm an attorney, but I don't specialize in product safety, so we hired specialized attorneys and a testing firm to make sure the Diaperbridge is approvable and complies with regulations and laws in the U.S."

The next step was to market the Diaperbridge. Garrett participated in three trade shows and got a handful of orders. "I'm normally a pencil pusher in an office, so I don't get to hawk wares to people. But at these shows, nobody got by my booth without getting their hand on the Diaperbridge. . . . It's exhilarating. I'll never forget that experience. I spent several days being a bit of a nut, making the most of every opportunity."

Garrett did get some sobering news while there. "A lot of the big guys said, 'Sorry, but you have to get this made in China; it's too expensive.' I don't want to do that. It's hard to coordinate. I have a full-time job as a pharmaceutical corporate lawyer, and my wife is the director of human resources at a hotel. We can't just take off for two weeks to do this. And besides, we don't have a distribution channel, such as Target, Babies "R" Us, or Wal-Mart, set up.

"We thought we could do it all and it would sell itself. From the stainless steel pins to the waterproof sticker, every step of the way has been a huge research project. I'm ready to stop getting my hands dirty with this business and be in a royalty situation. Lack of results might just be this product, or the way this particular business works. In pharmaceuticals, people come after the products." It seems there's a world of difference between diapers and drugs, he's finding out.

"We have a PR firm, but it takes a full, coordinated approach to sell a product—website, magazine ads, and a store presence—because people want to touch it. We went for broke trying to get on The Oprah Winfrey Show and news shows. But even when you're on for a couple of minutes, people don't buy your product; they just become familiar with it."

The Stackmans have invested a great deal trying to solve the world's diaper-changing dilemma. "Our costs of $200,000 have been spread over three years, and we have decent jobs, so it's not killing us." But then there are all those Diaperbridges languishing in their garage!

There's no question the Stackmans are pooped. "I'm a test-tube guy. I can develop the product, but I don't have the marketing ability or penetration strategies to actually sell it." Garrett wants someone to take his patent and bridge his efforts. But even though he's looking for some company to take over his product, he's not giving up. "I believe you've got to take a chance on something you believe in at some point in your life. It's been really fun."

Still, every time Garrett drives by the restaurant in Maryland, he stares at it and thinks, "You either did me a huge favor or you're mocking me, kicking me in the butt." Garrett will cross that Diaperbridge when he gets there.

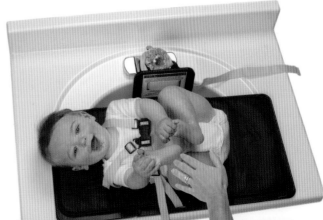

Baby on board

Re-Pillable Card®

For the Love of Aspirin

"All of a sudden, we had thousands of hits on our website and tens of thousands of orders."

"If aspirin, a pill that can open up your heart and thin your blood, were invented today, it would sell like crazy—even for five or ten dollars a pill, and even if it just got rid of your headache. But it's sold for two cents!"

After reading two articles about aspirin in *Men's Health* magazine, John Higgins was ready to put in his own two cents. One article touted aspirin as the greatest medication ever because it can stop a heart attack without damaging the heart. Two months later, a second article said that aspirin is so important to men's health that all men should have six bottles strategically placed in easy-to-reach spots hidden just about everywhere.

This struck John as ridiculous, so he wondered, "Where, on a man's body, could he carry a few aspirin to save his life?" In his pocket, they'd be covered in lint. Wrapped in foil, they'd be just one more thing to forget. Shoes, even penny loafers, were out. A hat? Nope, who wears a hat? And that's when it hit him— the wallet. Of course! Most men either carry or are close to their wallets 24/7.

So, what fits in a wallet?—money and credit cards. With that, John fashioned a prototype out of cardboard, like a thick credit card. He put it in the front slot but, when he closed the wallet, it got gigantic. Wallets are already too big. "My Irish grandmother in heaven must have guided my hands, because the next thing I did was put it in the top credit card slot. I turned away to take a call, and looked all over for the card. Then I tilted the wallet toward me and my mouth dropped open. I saw the proto-type in the fold. There's room—and enough for two!"

He called his best friend Ken Weinum with his bright idea and Ken, who had the money to back it, said "I'm in." A patent attorney's search came up empty-handed, and declared this invention too important to not be out there.

John showed and explained his crude prototype to an injection-molding company. It took a dozen tries on a CAD (computer aided design) program, but finally the design for the Re-Pillable Card was born. Like a credit card with a pillbox across the top, it can hold three pills in the left compartment and two on the right.

Because *Men's Health* magazine had inspired his invention, John wanted to share his product with them first. After seven months of trying, "We were in it. All of a sudden, we had thousands of hits on our website and tens of thousands of orders. We still get half a dozen a week from that article. People must be reading the old article in doctors' offices."

Three months later, *Diabetes* magazine ran an article. The medication for diabetics didn't fit in the original Re-Pillable Card, so at the suggestion of the editor, John created the Re-Pillable Max Card that fits in most wallets, but not

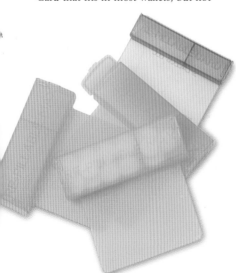

all. "Since it's so important to have your medicine with you at all times, if it doesn't fit in your wallet, buy a new one! Take the Max Card down to Macy's and try it out; make sure it fits."

Then John received an e-mail from South America, written all in caps. Someone had bought two, and loved them. Now he wanted to buy hundreds of thousands. John thought, "Yeah, right." A few months later, the same guy emailed to say he wanted 200,000 and where did John want the money transferred? John was still skeptical, until the funds hit his bank account a few hours later. His next e-mail showed what this customer wanted the 200,000 Re-Pillable Cards to look like: pewter gray, not John's blue, with an engraving of the Bayer logo and, above that, the Levitra name and flame logo. Between the logos was a tagline in Spanish.

These went to Mexico City for Bayer Mexico. "They could order millions now. I wanted it to be aspirin, but my first hit was for ED (erectile dysfunction) pills." John might still make it big with the aspirin makers soon. He is talking with Bayer Aspirin U.S. about plans to shrink-wrap the Re-Pillable Card as a value-added incentive to not buy generic.

The Re-Pillable Card is really catching on now, and each time a new chapter unfolds in the Re-Pillable story, it's told in the local Nashua, New Hampshire,

Keeping aspirin next to credit cards: makes cents.

newspaper, where John is a hometown hero. "Here's this little guy from New Hampshire exporting American-made products to Mexico. My congressman should be patting me on the back. That's how NAFTA is supposed to work, as a two-way street."

John has already sold well over 400,000 cards and was in the black after the first year. John and Ken's initial investment of $25,000 paid for the first mold and they've kept it rolling from the sales that came in. They now have four molds, so they can make 250,000 cards a week. "*Men's Health* was the key to everything: our initial sales, *Diabetes,* and Levitra. And we haven't even spent a nickel on advertising." At 59, this former print broker says he might retire yet. Not a bitter pill to swallow, eh?

Mother's Third Arm™

Armed and Ready

STAT BAR

PATENT: US #5823486 as Universal Flexible Arm; TM as Mother's Third Arm

PRODUCT PRICE: $19.95

STATE: Arizona

INVENTOR'S AGE: not saying

INVENTOR'S PROFESSION: homemaker

MONEY SPENT: $300,000

MONEY MADE: still in the red

WEB ADDRESS: 4innoventions.com

"Sales are great. I'm holding my own."

What mother hasn't wished for an extra set of hands, especially during those tough infant and toddler years? Margo Smith, mother of five and grandmother of eleven (blending families with her husband, she has a total of twenty-six grandchildren) knows this wish all too well.

When you're a mom in the trenches, being armed with a third arm could give you that extra hand to win the baby battle. And this grandmother wanted to help mothers everywhere.

Margo wanted to make something that would hold baby bottles, sippy cups, and toys, and would attach to car seats, strollers, tables, high chairs, and shopping carts. She went to stores and confirmed that nothing like this was in the marketplace. Next, she did the patent search, then the prototype construction, and finally did focus group testing to get the public's opinion. The focus groups loved her invention. Margo was ready to move forward.

Sometimes in product development, an invention takes on a life of its own. The inventor recognizes one user group for the product, but the public sees another user group. That's what happened with Mother's Third Arm. This baby product found its way into the arms of wheelchair users and their caretakers, who have embraced Mother's Third Arm and the independence it provides. If wheelchair-bound children can move their heads to where the cup is held, they can sip from a straw. Margo says, "A product is either a need or a want. For mothers who have a child in a wheelchair, Mother's Third Arm is a need." And they're grateful to have this need met.

Margo thought of the idea in 1996, got her patent two years later, and then had 10,000 manufactured in

Hands down, a smart idea

Phoenix. She brought her costs down by getting 20,000 made in China. Out of those 30,000, she only has 1,500 left. Mother's Third Arm is not sold nationally, but an earlier model was—through Toys "R" Us, Baby Depot, and other outlets. That one-size version broke when people tried to put different size cups in it. So Margo took it off the market and improved it to hold various sized bottles and cups. She has been selling it through catalogs, the Internet, and her office in Phoenix, as well as giving away thousands to children's disability charities.

Margo hopes that, before long, her invention will be in all the stores and on the Home Shopping Network. She is working with a big company that has plans to take Mother's Third Arm all over the U.S., Canada, and Europe. "I've

been looking for these people for a long time. Instead, they found me through a website." If the deal goes through, Margo will no longer be responsible for manufacturing Third Arm. Instead, the big company will make the product and the patent will remain in Margo's name. Margo says that, if it happens, she'll rest easy in the arms of this deal.

Last fall, Margo entered Mother's Third Arm in Proctor & Gamble's achievement contest, and out of 400 products, it came in third. Do you see the irony here? Mother's Third Arm came in third. And it turns out, the contest brought in more than just kudos. "Target is very interested, as is Canadian Tire."

To get Mother's Third Arm this far, it has cost Margo an arm and a leg. She's spent about $300,000 over nine years and is now more than eager to get out of

the red. "Sales are great. I'm holding my own. But I'm still putting my own money into it." She lost a bundle through scams. "I was scammed by three companies. As soon as I got my patent, a company said that for only $5,000 it would help me get my product out. Companies like these don't do anything for you but take your money. And it's too small an amount to sue over."

Through it all, she's had the unfailing support of her husband, family, and friends. "The first few years, everybody was happy for me. Now when I see friends, I wonder if they're afraid to ask, thinking, 'Oh dear, is it still going?' No one ever told me I was crazy. People

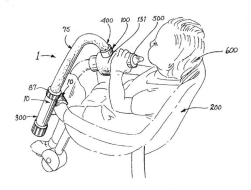

might have thought it, but they haven't said a word."

Whether Mother's Third Arm makes it or not, when you see the effort Margo has put into her invention, you just have to stop and give her a big hand.

Air Guardian™

Air Today, Gone Tomorrow?

"My son is now four years old and hasn't wheezed a day in his life!"

Hal Koch is one devoted dad. And his son Kyle is one lucky kid. Having suffered with asthma and coming from a family with respiratory problems including emphysema, Hal wanted to protect his son from developing breathing problems.

When Kyle was brought home from the hospital, the nurses said that he was not to go outside for the first month. Concerned about air quality, Hal surfed the Internet and learned that our air is not as clean as we think. Every year, over 10 million school days are missed and billions of dollars are spent on medical costs because of air pollution in this country. Hal learned that if a child's immune system is impaired early, it sets the stage for problems down the road.

And it is on the road that Hal focused on air issues. He wanted to protect his son's developing lungs. "People were having heart attacks from driving behind cars blowing exhaust in their faces. Every mode of transportation these days has an in-cabin air filtration system. The type of transportation that needs air filtration the most is the baby stroller."

Born and raised in Queens, New York, Hal saw parents using covers on their strollers even on nice days, worried about fumes from buses. Keep in mind that most kids sitting in strollers are breathing right at tailpipe level, so they are getting doused with diesel exhaust. Even away from the city, Hal saw many parents taking their kids to the school bus while a younger sibling waited in a stroller. These siblings would wave goodbye, then get a blast of bus exhaust blown right in their faces. Yuck.

So, along with customizing Kyle's stroller with, get this, mag wheels, headlights, taillights, undercarriage lights, rearview mirrors, DVD player, speedometer/odometer, temperature gauge, a cell phone holder and charger (because no parent should be caught without a cell phone), and a solar panel to keep everything charged— Hal invented the Air Guardian to filter the air Kyle breathed. "My son is now four years old and hasn't wheezed a day in his life!" I'm sure Hal breathes a sigh of relief over that.

Hal's first prototype came together in a day, with the help of his then two-year-old son, who put the O-rings on the tubing. Hal went to Home Depot, grabbed selected items off the shelves, and put them together. He named the first unit the Air Octopus because the ventilation tubes looked like an octopus's arms. Deciding this design was too big and bulky, Hal bought more things from Home Depot and made a few more pro-totypes. The result is the Air Guardian, which is about the size of three cassette tapes. It attaches to the inside of an enclosed stroller, car seat, or bike trailer. The fans in the Air Guardian draw

contaminated air through filters that, according to Hal, scrub the air clean up to 99.7 percent. The clean air is then blown toward the baby's face, like a breath of fresh air.

Hal says everybody in the whole world supports his invention. Hal took the Air Guardian to the Yankee Invention Expo and it proved to be the star of the show. A month later, he was on *The Tonight Show with Jay Leno* for his "Pitch to America." Kyle arrived at the show in his souped-up stroller. The audience got to vote—is the Air Guardian a likely "sell" or "no sell." They voted it "a sell." Hal's next big move was a spot on ABC's *American Inventor* show. He's gotten some impressive airtime.

Hal was motivated to take the Air Guardian to market after he lost his job as a limousine driver two years ago, but he has always landed on his feet. Now he's customizing strollers.

"The Air Guardian is not on the market yet, but it is patent pending. People come up to me wanting to be an investor on this project. But I have perfect credit and I'm not looking for just the money. I need someone who can help me with research and development,

someone who has the knowledge to bring it to market."

Hal has worked out most of the details; he even has a product motto: The Air Guardian: Bettering the Quality of Life for Our Little Ones.

With any luck, the right guardian angel will partner with Hal and together they'll protect the next generation's lungs. But right now, it's still up in the air.

OrthoTote™
Strapped for Cash

STAT BAR

PATENT: #669863

PRODUCT PRICE: $11.95

STATE: New York

INVENTOR'S AGE: 61

INVENTOR'S PROFESSION:
businessman

MONEY SPENT: $300,000–$600,000

MONEY MADE: undisclosed

WEB ADDRESS: orthotote.com

"It takes tremendous perseverance."

Every business traveler knows the drill. You're running through the airport looking for a gate, ticket clutched in your hand. As you weave in and around the other travelers, you can feel the strap of your travel bag slowly slipping off your shoulder. What a pain. My own solution would be to just work out more and have bigger shoulders. David Finkelstein had another idea.

David's shoulders also would not hold the strap of his travel bag. "Every five feet I walked," he remembers, "the bag fell off my shoulder. It was very annoying." This inconvenience spawned an invention, and ultimately a new career for David, a businessman who never dreamed that, at age 61, he'd be an inventor.

David graduated with a degree in accounting in 1967, but "I always hated it," he says. "I have a creative bent, and accounting is the opposite of that." So after three years in accounting, David went into the medical supply business, where he stayed for thirty-five years and was successful enough to eventually finance his invention.

Now, let's go back to the airport, where a frustrated Finkelstein fought to keep his bag on his shoulder. David considered the problem and realized he needed to reverse the power of gravity, which was pulling the bag down the slope of his shoulder. He devised an attachment to a shoulder strap that would reverse the downward angle of the shoulder. The OrthoTote was on its way to arriving at airports everywhere.

"I drew up pictures and went to an intellectual property attorney. This was before September 11, 2001. It took two years to get the patent back. I got a utility patent, which is better than a design patent."

David took his drawings to an engineering firm to produce prototypes, all financed out of his own pocket. "It took twenty-five changes. It was like Goldilocks

Shoulder a heavy load

and the Three Bears—this version was too soft, that version was too sticky."

David says he was "blown away" by how long the process took. He would wait six to eight weeks to receive a prototype from the engineering firm, only to see within minutes that the version was flawed. Then he'd have to wait several more weeks for the changes to be made.

Throughout the process, David's two daughters served as "test pilots" for OrthoTote, giving him frank feedback. He left his medical supply business to concentrate fully on the OrthoTote, and enlisted the partnership of his lifelong friend, Brian.

In mid-2005, test runs complete, the OrthoTote was finally ready for its debut. "I decided to manufacture it fully in the U.S. Although I probably could have saved money going out of the country, I believe that products made in the U.S. are the highest quality.

"I didn't have millions to spend on advertising," David explains, "so I went to the library and found a listing of all the magazine editors that I thought might have the slightest interest. I sent each one a letter, a sample, and a photo. As a result, the OrthoTote was written up in eight or ten magazines."

David Finkelstein is proud of his product. His goals for the future of the OrthoTote include developing special versions of the product for women's handbags and for backpacks, as well as getting into a contractual agreement with the post office to get an OrthoTote on their carriers' bags.

Looking back on the process, this grandfather of nine reflects, "The road-blocks along the way for an inventor are tremendous. The money, the work, the time it takes—it's all much more than I thought. It takes tremendous perseverance."

But it has also been tremendously ful-filling. For David, becoming an inventor has been the realization of a dream he didn't even know he had. In creating the OrthoTote, a product which he insists "won't change the world, only yours," David has indeed changed his own world.

No question that getting OrthoTote to market has been a haul, but luckily David's OrthoTote makes heavy tasks a little easier.

Take-Out-Time-Out® Mat

Because Bad Behavior Happens Away from Home, Too

STAT BAR

PATENT: information not provided

PRODUCT PRICE: $12.99

STATE: South Florida

INVENTOR'S AGE: 36

INVENTOR'S PROFESSION: full-time mom

MONEY SPENT: $120,000

MONEY MADE: has made some back, but still in the red

WEB ADDRESS: take-out-time-out.com

"Once the product hits the market, you're building a reputation. It took a lot of time, but it's paying off."

Being a mom is a full-time job, especially when you have two little ones. Lisa Bogart Carvajal knows parenting stress. This South Florida mother has two boys, now aged four and six. When her oldest hit the "terrible twos" and she had a baby in tow, Lisa had her hands full.

Lisa wanted to make sure she managed this stage of their lives without losing her mind—or losing control. After researching parenting books, she found that the "time-out" method was the most successful. You know the drill: Your kids do something bad and you give them a "time-out." Translation: they have to sit on a chair in their room for a few minutes as punishment for their crime.

But who stays home all day? And kids don't save their misbehavior for when they're at home. It didn't take long before Lisa realized that "time-outs" in public weren't working. Her kids saw the weakness in the system, too. It was as if they were thinking, "Ha, ha. I can do whatever I want." It got really ugly.

Not wanting to be beaten by a two-year-old, Lisa decided that she needed to provide a place for them to sit during time-out—wherever they were. So first she looked for a fold-up chair that she could carry around and use when they were out and about. She came up empty-handed. That's when she decided she needed to go to the mat for a solution. Her friends agreed and encouraged her to design and patent the very item she was shopping for.

Lisa worked on product development for eighteen months. After coming up with a mat design, she involved focus groups and tested different materials, sizes, and portability. With some materials, kids would slide the mat around. Not good. After all, time-out isn't sit-and-spin-around-on-the-floor time. Lisa's time-out mat could not be fun.

Lisa found what she was looking for in mouse pad material. The rubber bottom keeps it in place, and if it's left on the floor and someone steps on it, he or she won't slip and fall. The material folds easily and can be stuffed into a purse or bag. Also, it's waterproof, so if you're at the mall and the floor is dirty, or at the playground and the ground is wet—who cares! You've got a clean, dry spot for your child to sit. This material is incredibly durable. Lisa knows; she's been using the prototypes for three years. She washes them in the washing machine frequently and they still look brand new.

"The Take-Out-Time-Out (TOTO) has not only helped me when we're out, but it's helped me at home. I've been able to be consistent with the time-out spot because I'm able to move the spot. Now, if I'm cooking dinner and one of my boys shows unacceptable behavior, I just place the mat on the floor next to me in the kitchen. If I'm working on the computer, I place the mat next to me at my desk. I don't need to stop what I'm

doing. TOTO has simplified my life. My discipline method is consistent, and my kids' behavior has improved as a result. I keep one in my purse, one in my car, one upstairs, and another downstairs."

Lisa tried four different manufacturers to find a source that could make a good product quickly and efficiently. She wanted to keep it in the U.S. but it would have tripled her cost, so she's manufacturing the TOTO overseas. The amount she's invested almost makes her want to cry and, at six to eight hours a day, the mat business is a full-time job added to her already full-time mom job. "Ironically, though, without TOTO, I couldn't do it—it wouldn't work, or I'd be neglecting my parenting, which I don't want to do."

Lisa's sales background and entrepreneurial bloodline (her grandfather and father each started businesses) keep her going. She tested the TOTO with mom's groups when going through product development. She recognized this as a crucial step, saying, "Once the product hits the market, you're building a reputation. It took a lot of time, but it's paying off." Once she had the product, she focused on advertising and the press. Then, after getting the word out, she worked on distribution. Getting it into retail stores has taken some work because it's an entirely new product. So she's focusing on building

"Think about what you did."

the awareness. After that, she'll work on branding.

Over the years, competitors stole her idea and even lifted her website copy verbatim. Her husband, Joseph, warned her to be prepared, that others might imitate her idea. The best approach is to take every imitation as a compliment. Her job is to focus on keeping the lead. Lisa has also found that imitations help build awareness, educate consumers, and increase customer base—she's working hard to be sure it stays her customer base.

Lisa has put a great deal of time and money into her TOTO mats and she's not giving up. Lisa is in this for the long haul and there are no time-outs in sight.

Roadside Message Board™

Frustrated Inventor, Sign Here

STAT BAR

PATENT: US #6688027-B2
(Feb. 10, 2004)

PRODUCT PRICE: $9.95

STATE: New York/Massachusetts

INVENTOR'S AGE: 45

INVENTOR'S PROFESSION: service
station manager, tow truck driver

MONEY SPENT: $100,000

MONEY MADE: "may have made
about $10,000 back"

WEB ADDRESS: marketlaunchers.com

"I'm lucky if I make 50 cents per piece."

As a manager of an Exxon service station and a tow truck driver for fifteen years in New York, Fred Fink saw his share of cars abandoned on the side of the road. (Fred says AAA estimates 25 million breakdowns a year in the U.S.) Fred has spent countless hours talking with police officers who stopped at his station, trying to find the owners of abandoned cars.

In today's go-go world, stranded motorists no longer wait for a tow truck. They use their cell phones to call a family member or friend to pick them up and get them where they need to go. Most figure they'll take care of the car later. But they can't take care of a car that's gone. Police ticket or tow abandoned cars—unless the driver has left a compelling message.

Seven years ago, Fred invented the Roadside Message Board as a handy, portable way for stranded drivers to get their message out—before it's too late. It's a license-plate-sized board that fits in the glove box, under a seat, or over a visor. Its two suction cups hold it to a car window or on the dashboard. It comes with ten prewritten messages that can be displayed on the message board. It also comes with a dry erase marker so drivers can write their own message on the board. There's even an emergency flasher that can be clipped to the sign or worn as the carless motorist walks along the road.

The preprinted messages Fred chose are the most common ones he encountered as a tow truck driver. These phrases of desperation were often scribbled in crayon or lipstick on a napkin or on the back of a receipt. The messages read like titles from a series of roadside sleazy novels: Car Trouble, Battery Dead, Out of Gas, Flat Tire, Overheated, Went for Help, Please Do Not Tow, Please Do Not Ticket, Be Back in __ Minutes, and—in case someone is just looking to get rid of the car—For Sale.

Fred knew the secret to success would be promotion. To keep costs down, he had his product manufactured in China, spending $15,000 for a mold and placing a large order of 10,000 units. The product arrived from China in pieces; then Fred hired companies that employ handicapped people to do the assembly work.

His first prototype was a flasher with a belt clip. Then he glued the flasher onto the board. But the handicapped workers weren't allowed to use glue because it might impair them. So he went back to his drawing board and came up with the solution. Fred modified the board to hold a clip-on light. That was just the ticket to keep his customers from getting a ticket.

At first, his family and friends were very excited, but didn't realize how much it would cost to proceed. Overall, he has spent $100,000 in the last seven years. It

Just about all car calamities are covered.

took five years to receive the patent at a cost of $10,000. (He was told it would cost $5,000, but it got rejected the first time and had to be resubmitted.) The mold cost $15,000. Liability insurance costs $2,000 a year.

With the bills adding up, Fred recently decided this venture was experiencing its own emergency. "I'm lucky if I make 50 cents per piece," he said. After paying for the patent, making the mold, manufacturing the product, traveling to conventions, incorporating, and keeping liability insurance, there's not much left in profit for this idea man. So after selling 9,200 of the 10,000 units he had made, Fred closed his corporation because it took too much money to keep

the business running. If he wanted to promote the Roadside Message Board, he'd have to spend more money to travel and hire salespeople.

At this point, Fred is waiting for people to contact him. He thinks it would be a great giveaway for car insurance companies. And the board provides a space to promote a product, leave a message on a store window, or occupy kids in the car with a game of tic-tac-toe. He's looking to sell the remaining Roadside Message Boards. He has them on a website, MarketLaunchers.com for $9.95. His dream is that a company will buy the mold and give him royalties. He's ready to walk away from it—much like the motorists he's tried to help.

Walk-O-Long™

A Step in the Right Direction

Inventors can be inspired by anything. Usually, a problem that begs for a solution gets an inventor thinking. For Jeff Zinger, necessity was truly the mother of invention. He had just undergone back surgery, but his 10-month-old daughter, Faith, didn't understand that. She wanted to pull herself up and walk. Faith really didn't care that her dad couldn't keep bending over to help her. What was Jeff going to do?

He saw Faith's nanny use a towel to hold Faith and Jeff thought there has to be a better, safer way. The former plumber went to a fabric store and made the prototype for the Walk-O-Long. His first prototype was his last. It worked.

The Walk-O-Long is a spongy fabric tube that fits around a child's chest and under his/her arms. It allows a parent to stand up tall and still have a firm grasp on the child. In fact, Jeff says that he used the prototype Walk-O-Long for thirty to sixty minutes at a time, and in about five days, Faith learned to walk. As people started to see Jeff and Faith using the Walk-O-Long at restaurants, Disneyland, shopping malls, and grocery stores, he would get questions about where they could buy one.

Ding! The lightbulb was on. Jeff realized that his Walk-O-Long might not only help his daughter walk; it

For Jeff Zinger, necessity was truly the mother of invention.

THE WALK-O-LONG™
Balance   Coordination   Self Confidence

could also help this plumber with a back problem take the first steps toward a new career—and get him back working. So he started the process of filing the patent paperwork for the Walk-O-Long. His parents and his wife's parents were very supportive. Even though his brother and sisters and his wife's brother and sisters made fun of the idea, that didn't stop him.

Jeff spent the next year working on packaging and advertising. He thought it was only appropriate that he use his daughter's face as a logo. After all, it was because of little Faith that the Walk-O-Long was invented.

In its first four months in stores, Jeff tells me he sold about 2,000 Walk-O-Longs. They sell for about $25 each, so you do the math. Despite the sales, Jeff says he is still in the red.

Once the Walk-O-Long got placement in stores, a funny thing began to happen. Jeff found his product had more uses than he could have imagined. Parents could use it to help their children down a playground slide; it could help a child get used to being on ice skates; it could even help when caring for kids with special needs. Recently, Jeff has been in talks with child disability educators at the University of California, Irvine, Children's Hospital of Orange County, the Foundation for the Junior Blind of America, and many parents of children with cerebral palsy.

Who would have guessed that material wrapped around a foam tube could be so handy? I guess you just have to have a "little Faith."

Taking a toddler for a walk

Umbrella Article Holder™

Ready for a Rainy Day?

"I come up with three new inventions a month."

In his 70s, Clarence Thomas (not the Supreme Court Justice) is a self-labeled "master" inventor. His sister calls him a "fanatic junk man." His patent office must call him a "regular" as he has 235 patents for all kinds of gadgets. That's right, 235 patents. Clarence says that people love them, but won't buy them. He was in the patent office with a germ catcher—"whatever you touch, it's covered with germs"—when I caught up with him.

"My attic, garage, and house are filled with these things. I'd like to sell at least one. I come up with three new inventions a month." The Umbrella Article Holder is one such idea that came to him on the way to his job as a building maintenance supervisor on Wall Street. He likes to read the newspaper while walking down the sidewalks of New York City. The weather isn't always sunny, so he often carries an umbrella along with the newspaper. He found it inconvenient to hold the paper with one hand and the umbrella with the other. So he decided to make something that would serve as an extra hand. Who hasn't wished for an extra hand from time to time?

The Umbrella Article Holder has a strap, similar to a blood pressure cuff, that fits around either leg and lets you insert an umbrella so it stands straight up. After he created that, he decided something else should be added, so he created a device for the handle of the umbrella. It's a square that covers half of your arm with a compartment to hold a coffee cup or a soda. Drinking his coffee, he decided to put a cigarette pack holder on it. And, of course, that led to needing a cigarette lighter near the pack, and for some reason a ballpoint pen. The pen might have led him to think of writing lists, because the next addition was a device that attaches to the arm square to hold a three-pound bag of groceries. And if you're shopping for groceries, you

It's not just for umbrellas.

tant—$152 for each unit. "I'd have to sell it for over $300. Who's going to buy it for $300? Nobody. Unless I was a celebrity." Then he went to the Yankee Invention Expo in Connecticut and met someone who said he could make it in China for under two dollars. "I couldn't believe it. I thought he was playing a trick on me.

"I've lost over $50,000 on the last ten of my inventions. An innovator in Wisconsin put one on the Internet for two years. He was going to sell the product, but didn't get a single call. I found out that it's not worth it and decided that instead of playing it that way, I'd try this guy and spend my money in China. So I sent him one of the seven prototypes I had made. It took several months; I just got it a month ago . . . I had spent $800 to have Made in America labels made—now I can't use those. I'll be fined if I do. I'll have to put the American labels aside and have the new manufacturer make a label that says Made in China."

Clarence is now ready to mass-manufacture his Umbrella Article Holder. "At my age, I can't keep playing around. I must get things moving or my ideas won't get out. My wife

might wear a coat—so he added a device that will hold a short coat, like a suit coat. And if you need a coat, it might be because it's raining, so you need an umbrella . . . you get the idea.

It took Clarence three tries to get the first version to work with a curved-handled umbrella, making it fit without throwing the user off balance. Then he had to make a different version to fit straight-handled umbrellas.

Clarence had a factory in New York City make seven prototypes. They came out beautifully, but the cost was exorbi-

wouldn't know what to do with these inventions if I died. I'm getting frustrated. I've spent money for a number of years, and had no return. I just want to sell one of my inventions before I leave this world. Get paid and move on. That's what I'm wishing for."

Clarence believes in his Umbrella Article Holder. After all, it gives users a free hand. You've got to hand it to this inventor; he is the real article.

Shoulder Dolly®

Hello, Dolly

STAT BAR

PATENT: US #6729511

PRODUCT PRICE: $59.99 for Light Duty, $300.00 for Heavy Duty

STATE: Washington

INVENTOR'S AGE: 29

INVENTOR'S PROFESSION: business owner

MONEY SPENT: not saying

MONEY MADE: not saying

WEB ADDRESS: shoulderdolly.com

"I needed investors, so I moved to where people with money live— Aspen, Colorado."

They tell us we should all lift with our legs, not our backs. Thomas Dent III found a way to lift with his brains.

This 29-year-old entrepreneur financed his college education, two degrees in economics and sociology, by moving appliances. While on the job, he decided there had to be a way to make carrying heavy appliances a little less backbreaking.

Thomas experimented with straps going under the appliance to allow two movers to lift it while in an upright position. "This way, the larger shoulder and leg muscles lift the weight, decreasing strain on the lower back, hands, biceps, and forearms. Moving large objects becomes easier and safer. Plus, the hands are freed up to guide the appliance rather than hold it."

The strap system worked. Thomas named it the Shoulder Dolly and used it at his job to lighten the load, making heavy appliances much more mover-friendly. His system even allows a 100-pound woman to lift a full-sized refrigerator. OK, put down the refrigerator. We get the idea.

When it was time for grad school, Thomas put his straps away to pursue a master's degree in international economics down under. "I only lasted five months in Australia. I just couldn't get this idea about developing the Shoulder Dolly out of my head. My dad thought I was crazy to come back to the U.S. and chase what he thought was a get-rich-quick idea. But I had to. And once he understood what I was doing with it, he became very supportive."

At age 23 and carrying $40,000 in student loans, the inventor of the Shoulder

Dolly was strapped for cash. "I needed investors, so I moved to where people with money live—Aspen, Colorado. I even slept in a tent the first summer I was there, before I could get established."

Once again Thomas got a job doing what he knows best—moving appliances. He started to work for a high-end appliance retailer Contract Appliance Center in Glenwood Springs, about forty miles from Aspen. Naturally, he used his Shoulder Dolly prototype on every haul. Not only did the shop owners, Tom and JoAn Knipping, love the Shoulder Dolly, they became backers of Thomas and his invention. David Cook, editor of the *Aspen News*, also got interested and so did Craig Wilkening, an account executive with an appliance manufacturer. In 2001, they formed TDT Moving Systems, Inc. to launch Shoulder Dolly.

"I put in a lot of effort applying for a provisional patent. I did the description and drawings, and wrote it up as perfectly as I could. Once I got these investors interested, I hired a patent lawyer to apply for a utility patent." That was the first hurdle Thomas has had to shoulder.

Next, he started participating in hardware trade shows—dozens of them—across the U.S. and Canada. People were impressed with his Heavy Duty Shoulder Dolly, but priced at $300 each, he couldn't sell a large quantity of them. "No matter how marvelous and efficient

Portable TV?

they are, I had few repeat sales. Because the Shoulder Dolly is so strong and durable, they don't wear out."

Thomas realized he needed a Shoulder Dolly designed for a wider range of consumers at a lower price point. "In effect, I made a knockoff of my own product. I call it Light Duty." This lighter-weight product is manufactured in China and sells for $59.99 through hardware retailers in North America.

The biggest and best exposure happened when Light Duty debuted on QVC shopping network. Thomas now plans to create and air infomercials, starting in small-city markets. He'll test-market them before spending larger sums to run infomercials in larger metropolitan markets.

Early marketing, publicity, catalog, and website exposure have resulted in selling 20,000 units of Light Duty since it was introduced in 2003. "We have fifty distributors for the Light Duty version in just two-and-a-half years, plus it's sold through Northern Tool and Harbor Light catalogs."

TDT Moving Systems is now headquartered in Vancouver, Washington, where Thomas and his girlfriend Brenda Castine, who works full-time in the business, moved to be closer to his family. The growing company now seeks investors to go international. "We do have distributors in other countries and are looking to file patents in many of them." It looks like Thomas Dent III has found the Shoulder Dolly to be not only an uplifting and moving experience, but also a weighty career.

EasyDown™

Easy Down, for Those Hard Up for a Way Out

STAT BAR

PATENT: pending

PRODUCT PRICE: $900 for automatic version, $300 for manual version

STATE: Massachusetts

INVENTOR'S AGE: 71

INVENTOR'S PROFESSION: product designer

MONEY SPENT: $20,000

MONEY MADE: $0

WEB ADDRESS: easydown.com

"A panicky person with no training should be able to use it safely."

Did you know that fire departments' ladders can only go up to six floors—and only from a side street—and only if the fire trucks get there in time?

After the tragedy of 9/11, it's no surprise that a colleague of Herb Loeffler's, Ivars Avots, recognized the need for a means of escape from tall buildings if the normal exits, such as stairs and elevators, aren't available. Looking through newspaper articles about tragedies, however, he discovered that the need for an escape route isn't a rare occurrence. People get trapped not only in 100-story buildings, but also in eight-story buildings. Ivars had a vague idea for a solution, but didn't have the technical background to make it work. He needed an engineer's brain and an inventor's heart to take this leap with him. He found that in co-worker Herb Loeffler.

Both men worked for a Boston-based industrial research company before the company closed its doors. Herb, an MIT graduate in mechanical engineering who

also has a degree in industrial design, now a 71-year-old, semiretired product designer, became the brains of the operation. The project was funded by the idea man and another colleague of the Boston firm.

Concluding that a market exists for an individual "descender" device that required only minimal skill to operate, they worked off a rappelling model that mountain climbers use. But while mountain climbers are trained to manage the speed of descent, the average person isn't. Another issue: rope is heavy. One thousand feet of rope is strong enough to hold the weight of a single person, but weighs more than a person can lift. A thousand feet of cable has the same strength as 100 feet of rope. So they went with cable.

The next step was deciding what to put the cable on. Mountain climbers throw their ropes over the side of the cliff and use a device to slide down. That's not practical for this use. Herb explains, "The cable needed to be on a reel with speed control—something that could sense the speed and apply the right amount of friction so the thing couldn't run away with you. A panicky person with no training should be able to use it safely."

As a product designer, Herb kept it simple. For the automatic model, he used a centrifugal clutch, as in snow blowers and chainsaws, to provide the speed control necessary. When the user goes faster, the clutch puts on the break harder. For the manual model, Herb added a handbrake for starting and stopping. A knob releases the cable. As you crank in one direction, it lowers you down; if you stop cranking, it stops

moving. You aren't actually cranking your own weight; you're just releasing a clutch. Because you can crank only so fast, the handbrake provides speed control. The crank also allows you to get used to how the harness feels while dangling out of the window before letting go of the brake. Then you can go down gently. Easy does it. Hence the product name—EasyDown.

Just the thought of having to use a product like this one has me shaking. But I guess if I were trapped, I would learn to love my EasyDown.

The manual model comes in at a third of the cost and half of the weight of Herb's automatic model. "The manual model is the device of choice from five to ten floors. Any higher and you'd want the automatic model. No one would want to crank that far in an emergency. And with its lower price, the manual model is where the market would go. But our biggest concern is that people don't want to think about safety."

It's a serious concern—after all, we humans don't like to think about our own deaths, much less prepare for them. But even if individuals don't want to contemplate mortality, companies do. Herb sees a potential market with companies that sell safety equipment to firefighters or miners, for example.

Easy does it with EasyDown

The team applied for a patent two years ago and have had some action on it. They haven't gone into production, but a partially completed design proves they can manufacture them at a moderate cost. To make it cost-effective, they envision producing 5,000 units. At 100,000 units, the price would be cut in half.

The fact that we might ever need an escape product can be depressing. But the probability that it might save our life, well, that's the upside to EasyDown.

Pick Up Hoops
Sportbinox
Stout's BackSaver Grip
Going-Going Crazy Game
Trangleball
shootAndstar Rebounder
SwingSrite
Pohol
Scope Mates
Boomwhackers
Pulse Clock
HanuClaus Hat
Harness Play Pack / Picnic Party Cloths
Panic Mouse

At Play

We work hard and we play hard. Americans are very proud of how seriously we play. We tell our kids that games are all about fun, but for most of us, games are all about winning. Inventors know we have a passion for play. They know we'll spend our hard-earned cash to make that playtime even better.

So while the rest of us are busy playing our games, inventors are busy trying to make those games easier, more accessible, and more interesting. Whether it is being able to watch sports with hands-free binoculars or turning a pickup truck into a basketball court, these inventors have made recreation their playing field.

You'll meet some inventors who have come up with entirely new games. Will Trangleball® replace volleyball? Will you go crazy over the Going-Going Crazy® Game? I found an inventor who is trying to make fishing easier and still another inventor who claims to have made a good old-fashioned turkey shoot so simple that it is almost child's play. There's even an inventor who has found a way to mix Christmas and Hanukkah. Inventors go where angels fear to tread.

If you are serious about your playtime, then get ready to meet some men and women who are serious about their playtime inventions.

Pick Up Hoops™

He Shoots, He Scores

*"...it can be set up in thirty seconds
without tools and the whole system
weighs less than 60 pounds."*

To understand 40-year-old Jason Parr, you need to know that this guy was a gym rat. He lived to play basketball. He described himself to me as a "hoop head."

When Jason wasn't playing basketball, he earned his living working as a coordinator for an off-site data protection company called Iron Mountain. But even so, Jason was still a "hoop head." He and his co-workers always tried to get together after work to play some hoops, but going to a nearby indoor or outdoor court meant complications. Why not a basketball hoop in the parking lot at work? Great idea, but the boss wasn't crazy about it. But if they wanted to pay for the hoop and put it in, it was fine by him.

Jason thought about it, including getting "one of those so-called portable types to take to and from work." Then Jason Parr got an idea that would forever change his life.

He happened to drive a pickup at that time. What could be more perfect than to have a hoop made to go in the back of his truck? He could have a basketball court with him anywhere he went. He could play anytime anywhere he wanted. Sweet.

Jason's pickup truck was already three feet off the ground, so he just needed seven more feet to meet the regulation height of ten feet. He sketched something that resembled scaffolding in the back of a truck that had a backboard hanging on it, then passed the ball to his uncle, Jon VarnHagen. Uncle Jon was an automotive engineer for Chrysler Corporation. Jon, who also loved basketball, immediately saw the genius behind a mobile instant basketball court.

Jon approached the situation from an engineer's perspective, establishing design requirements up front. First, the backboard and goal must extend far enough away from the back of the truck to allow adequate baseline play. Second, the system must able to be set up or taken down by one person, without any tools, in under a minute—otherwise it might be considered too much trouble for most consumers. Third, the design would need to fit all truck sizes: short or long boxes, full-size or mini pickups, and 2- or 4-wheel drive. Fourth, the hoops would need to be easily removable from the truck when the truck bed was needed for uses other than basketball (like that would ever happen).

The design that came to Jon was like a convertible top on a car—after all, he did work for Chrysler. The hoop system would need to fit neatly and compactly in the space available, which is the bed, then extend up and out, away from the back of the truck, in the play position. The resultant design achieved all the design requirements and then some. Rear overhang of the backboard extends about four feet from the truck's tailgate. The hoop angle can be adjusted to compensate for parking on a sloped lot. What's more, it can be set up in thirty seconds without tools and the whole system weighs less than 60 pounds.

They named the product Pick Up Hoops and made a few prototypes. Jason brought it to work and his fellow co-workers loved it. He showed the Pick Up Hoops to everyone he saw playing basketball and they all loved it. One group that was especially excited about Pick Up Hoops was the Michigan Wheelchair Basketball Team. You see, most public basketball courts are a slab of asphalt with grass or dirt around the court. For basketball players in wheelchairs, rolling off the court onto the grass can be a real problem. But with Pick Up Hoops you can set up your court anywhere in the middle of a parking lot and play b-ball. In town parades, Jason drives his pickup truck down the parade route with the Michigan Wheelchair Basketball Team playing right behind. Cool, huh?

As for Pick Up Hoops, Jason is not trying to sell individual units. He wants some large company to license the concept from him. He has knocked on every door he could think of trying to get some company interested, including Spalding and just about every backboard company. He's talked to all the American car manufacturers and now he's planning to hit on foreign carmakers. So far, nobody wants to play ball.

"When I was growing up, the only way you had your own hoop was to have a garage. But now with our product, if you have a truck you have a truly portable hoop game."

Jason is quick to point out that there are other portable hoops, but says they all feel like toys. His hoop has a professional backboard of 54 inches, whereas most portables are only 42 inches. And, most importantly, his hoops can take a beating and even handle a dunk.

Jason can't see why he hasn't found the right company to license his product. He says one out of every seven vehicles sold in the U.S. is a truck. Young men love trucks and basketball; it just makes sense.

Jason's not giving up. When it comes to basketball, he's a competitor and this game is far from over.

This inventor's got game.

Sportbinox™

Driven to Invent?

Picture this. You're at a football game in the nosebleed seats. The action on the field is far far away. So you grab your binoculars and hold them up to your eyes. Here's the problem. How do you reach down to get your nachos and soda while holding the binoculars up to your eyes? It's impossible.

"So I thought, why don't I mount the binoculars on my head?"

That's the problem Tristram Himmele decided to tackle. Tristram is a 38-year-old who used to race cars in his twenties. As an avid sports fan, he knows firsthand that your hands aren't free while holding binoculars.

The search to solve this problem led the former racecar driver down a new road in life—the road to being an inventor. "I was tired of holding my binoculars and having to put down my Coke. So I thought, why don't I mount the binoculars on my head?"

Tristram's invention, the Sportbinox, is a hands-free binocular set with built-in AM/FM radio. It allows fans in even the worst seats in the stadium to get a great view of the game while keeping their hands free.

The road to developing this fan's Sportbinox has been a long and bumpy one with many pitfalls, roadblocks, stops, and starts along the way. Reflecting on the path behind him, Tristram says, "I've grown a lot as a person." First, the product went through several incarnations before the final version was created. In fact, an early version of the Sportbinox was mounted in a tray.

Once the Sportbinox design was complete, Tristram wanted to hit the gas, but instead needed to slow down due to the numerous speed bumps on the road toward manufacturing the product. One turn went directly into a dead end when he traveled to Hong Kong to meet with a manufacturing company. This company ended up, as Tristram says, "taking advantage of me," promising to develop the Sportbinox but then filing for a patent on its own.

The process also cost a lot more than Tristram had ever imagined it would.

"I've been able to finance it because of some good investments I've made in South Florida real estate." But even so, he adds, "I'm quite far in debt."

Another hard lesson to learn was how difficult it can be for an independent inventor to get his product into stores. "Buyers for stores don't want to take a risk with an independent."

Behind every great inventor there's a lawyer. For Tristram, that lawyer happens to also be his girlfriend. Her unfailing support and the fact that she's also a corporate attorney have been a big help.

Tristram's vision for the future of the Sportbinox is bright and clear. After all,

as Tristram sees it, "Everyone has two eyes," and therefore everyone can use a Sportbinox.

Tristram also sees the potential in renting the Sportbinox at stadiums, allowing people to spend a lesser amount to use the product without laying out the full price to purchase one. Tristram's plan for the road ahead is to use the public relations potential of the Sportbinox to get attention without spending dollars on advertising.

"The Sportbinox is highly visible because it sits on top of your head." So Tristram plans to wear the Sportbinox while sitting in high-profile seats at sporting events. By doing this, he hopes to attract attention and score interviews with reporters from large media outlets.

Eight or nine years after conceiving the idea for the Sportbinox, Tristram is now somewhat of an expert on what it takes to succeed with an invention. "It's been a learning process; I've learned so much about human nature. I am now able to help other inventors understand what it

takes. It's so important to have a large market for your product. And to make sure your product is a necessity, especially as economies are tightening."

With Sportbinox, this former racecar driver sees the road ahead and certainly has the drive to make it all the way on the road to success.

You see better and look great.

Stout's BackSaver Grip®

The One that Got Away . . . Is Back

"I have a ton of money in it—$250,000—and every dollar it has made has gone back into the venture."

To those of us who don't do it, fishing looks easy, maybe even boring. But those who fish would strongly disagree.

Imagine fishing for halibut, a 300- to 400-pound bottom-feeder that can measure 4 feet wide by 8 feet long. It might take an hour of cranking the fishing pole to bring it up from the bottom, plus an hour of trying to hold the pole against the strong twists of a fish that's bigger than you. After ten minutes, your hand is dead. Now what? Well, that's where Stout's BackSaver Grip comes in.

Ron Stout, an Alaskan diesel mechanic and avid deep-sea fisherman, once tore off a big chunk of his thumb while reeling in one of these babies because he couldn't hold the rod when it twisted. He says he bled more than the fish! Pretty image, huh? But now Ron has more than a fish story to tell. He's invented a solution and has three patents to prove it.

It all started when Ron and a charter captain were talking about how difficult it is to hold a fishing pole against the forces of an angry fish. They came up with a device to help hold the pole. Ron got scrap aluminum from a boat builder and made the first prototype. He attached this aluminum handle to an old halibut rod with Velcro, hooked it up to an engine, and nearly cracked the fiberglass rod with the incredible vertical lift. It worked.

In 1998, Ron hired Cadillac Plastic to build a wooden mold and form a plastic version: black with a clear T-handle; then he got a patent. Although the charter captain didn't put one dime into the venture, he wanted half of the profits. It was clear that this partnership wasn't going to sail into the sunset together. While the captain's name was on the first patent, Ron changed the design twice since and repatented it—in his name only. He has two different

The wrong way: brace for back pain.

trademarks in the U.S. and is waiting for the final patents to come from Australia, South Korea, and the European Union.

The BackSaver Grip attaches to the top of a fishing pole so you can keep your hand palm down. This position allows you to use your body weight and legs more than your wrists and forearms to reel in your catch. And on a charter boat, deckhands can reach over your shoulder and grab the handle to help you. Then, once fall and winter arrived, Ron discovered other uses for the BackSaver Grip. If you attach it to a rake or snow shovel, you can stand upright while using these tools, saving your back (that's where the name came from). It is so easy to attach, Ron has put it on every tool he owns.

"It's a cool invention if it will ever take. I have a ton of money in it—$250,000—and every dollar it has made has gone back into the venture. I received an inheritance when my dad passed away. That's the only way I was able to do this. Instead of building a new house—I live in a shack—I put it into this. I refinanced my house last year to change the mold and packaging, and get the overseas patents. It just has to take off."

In 2000, Ron got it in Alaska's Home Depot stores through the Buy Alaska program. In the first 110 days, he sold 650 grips. He wrote a letter thanking them and Home Depot teased him with promises of going national, but it never materialized. But today, the BackSaver Grip is sold in Ace Hardware, Do it Best hardware stores, True Value stores, and selected stores in various states, as well as on his website.

Friends and family members have supported Ron 100 percent. During the early years, his inner circle helped him drill, decal, and package the grips in his garage. Now, commercially packaged and prepared, they're poised to sell well. The grips have been getting a lot of attention thanks to two guys, David Wilk and Bob Franklin, who do all the traveling, marketing, and distributing now.

Dave and Bob got Stout's BackSaver Grip on NBC's *Today Show* on May 17, 2005, as part of a report on ergonomic tools to make gardening easier. They also got the BackSaver on QVC, although the experience ended up being disappointing. Dave and Bob worked with QVC to write the script and it was slated for an early summer show on a Saturday morning. At the last minute, QVC changed the segment from a garden show to one for people with degenerative diseases. The BackSaver was shown with a back massage chair, an eating bib, a neck support/stretcher, a bed pad, and a rake. Talk about fishing with the wrong bait! Ron's Grip sold only 1,000 units in four minutes, really bad numbers for QVC, and the shopping network sent back 7,000.

These days, Ron has backed away from the day-to-day operations. He returned to Anchorage and opened up a diesel repair shop before he got to the point of declaring bankruptcy. "I've put so much into it that I get frustrated with the disappointments. I've got too many years and too much money invested. I just want to hear when the good news comes in." Ron Stout is one fisherman who's waiting to catch the Big One.

The right way: back saved.

Going-Going Crazy® Game
Playing for Keeps

"Everything from the Going-Going Crazy name to the colors and design came to me in the dream."

Kurt Kirckof says he is all about his kids. When he goes to bed, his kids are on his mind, and when he wakes up, his kids are on his mind. And it's no wonder—he and his wife, Celestie, have five of them: Ashley 13, Aaron 11, Josie 8, Macie 5, and Emmit 2.

One particular night, Kurt went to bed and dreamed about a kids' game. "In my dream, I was playing this game with my two kids (Ashley and Aaron, who were then 8 and 6). I'd never played games with my kids. It's not what I grew up doing." Milking cows morning and night on a dairy farm, young Kurt had no time for horsing around. But now this father of five dreamed of a game called Going-Going Crazy.

"Everything from the Going-Going Crazy name to the colors and design came to me in the dream. When I woke up, I drew it, then put the drawing in my dresser drawer." A year later, Kurt pulled out his notes, bought a board game, and used the board and box as forms. "At the time, I did lettering for sides of vehicles, so I had vinyl and a special cutter on hand. I designed the game on the computer and, using the vinyl, laid it out on the board. For the box, I had a picture of Ashley and Aaron put on a big sticker that I wrapped around it." He played the game with his kids and their friends, and they loved it!

The next year, Kurt took his Going-Going Crazy Game to the world's largest invention show in Pittsburgh. "I went with the game under my arm and bought helium balloons from Wal-Mart to put on the corner of my table. Out of entrants from thirty-five countries from around the world, I got the gold medal! I left that place in tears. I knew I wanted to get this game made."

But getting a product made can be the ultimate game, and Kurt had no idea how to play. He had no information, no experience, and no contacts. When a guy who works with overseas manufacturing finally helped him get pricing in China to compare it to U.S. pricing, Kurt liked the numbers. He ordered 6,000, selling his very first game to a little store in a town thirty miles from his Minnesota home. For the next ten days, leading up to Christmas Eve, Kurt went on seven radio shows to promote it. "In those ten days, I sold 600 games out of two little stores in a town of 10,000 people. People were standing in line to buy Going-Going Crazy Game!"

Based on feedback, Kurt changed the game's original red, white, and black colors, to blue, yellow, orange, and purple—colors designed to keep kids interested—and repackaged it. Two years after winning his first-place award, he went back to the Pittsburgh show and got *another* gold medal. Since then, he's won so many first-place awards, he can't keep track of them all. His latest improvement, making all the pieces, including the board and spinner, glow in the dark, got three more catalog companies to pick up the game. He was on the cover of *Inventors' Digest* magazine for two months and has been in twenty different magazines and 500 different newspapers. But Kurt says that all that publicity didn't greatly

For this inventor, Going Crazy is a dream come true.

influence sales, since his product isn't available nationwide yet.

The Going-Going Crazy Game and related products are available on the game's website, but Kurt tries to route buyers to specialty stores that carry it in twenty-one states, Canada, and Japan. "Three years ago, Wal-Mart put me in forty-eight stores for a trial. We were the number one game for two years straight, but Wal-Mart won't put me nationwide unless I'm a million-dollar company. I wish I had a millionaire investor to back me. Still, that placement helped me build a sales history."

Over seven years, Kurt has invested $650,000 in real money, not play money. This inventor's playing for keeps! "If this game ever went nationwide, it'd be worth multi-multi-multi-millions overnight."

This venture has been tough for this 38-year-old auto body shop owner and

his wife, a registered nurse. "We're making monthly payments and are trying to raise a family. There's a point where you draw the line and say 'enough,' then pull the plug. But we just had more catalogs pick it up. I'm confident that it will work—you aren't going to sell 30,000 games if people don't like it. Besides, I'm at the point now where it has to work. I've got a lot of money invested! It's all a matter of getting the name out."

Target recently called Kurt to set up an appointment, and he has a good feeling about it. "If people will give me five minutes of their time, I can almost guarantee they'll walk away with this game. If you get one large retailer to pick up your product, then they all will. Once it's nationwide, we can do a big advertising campaign."

This game is far from over.

Trangleball®

Having a Ball

I don't know who invented baseball, basketball, or football. I'm sure the information is out there; I just don't know it. But I do know who invented Trangleball—New Yorker Mark Miller.

"Prototype it. It's difficult to build on an idea without something to physically touch and play with."

Mark says he doesn't care about his own fame and fortune. He just wants the name Trangleball to become known throughout the land. By that, he means he wants kids to play his game on the streets, at summer camp, and in gym classes, plus on beaches and places where people get together for fun. "Trangleball provides a lot of exercise in a little space and it's fun like you wouldn't believe."

Mark owned a music studio. One morning, waiting for a band to finish rehearsing, he picked up a ball and threw it at the corner of the room where two walls meet the ceiling. After throwing it at what seemed like the same spot, he noticed that the ball rebounded in a different direction each time. Sometimes it went left, some-times right, sometimes high or low. Mark thought that the unpredictable rebound was fun and could be used to improve ball players' reflexes.

Deciding to make a portable corner, he cut a 4-foot square of plywood into triangles, from corner to corner "the way my mom used to cut my sandwiches." He then nailed three of the four triangles together to create a corner. Mark had his portable corner, and it worked just fine.

Soon after he created it, he flipped it over and realized that his corner was a three-sided pyramid. He began to throw the ball at one of the panels, but the ball rebounded right back to him. Boring. Then Mark had an "aha" moment. "If I ran five steps to the right of the panel

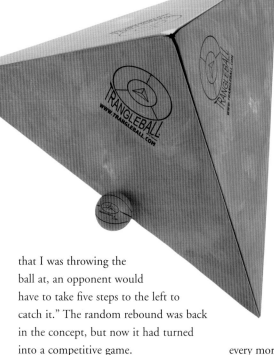

that I was throwing the ball at, an opponent would have to take five steps to the left to catch it." The random rebound was back in the concept, but now it had turned into a competitive game.

After a few hours of playing this corner game with his employees, Mark realized that the other two sides of the pyramid weren't being used, so he had them play three games at once: one against each panel, for a total of three balls and six players. As time went on, he eliminated two balls and created a passing game between the players. That's how his inverted corner became Trangleball.

Mark's advice to other inventors: "Prototype it. It's difficult to build on an idea without something to physically touch and play with. A prototype will build fresh new ideas. Without my first proto, I'd still be throwing the ball at the inside corners instead of flipping it on the pyramid side."

Today, Trangleball has a circular court (portable), a yard-high pyramid in the center of the court, and a squishy ball the size of a baseball. "I even created the Trangleball logo in the shape of the court itself."

If you go to Fire Island's Ocean Beach and walk along the shore, you will see six players bouncing a ball off one of the three faces of the pyramid in the center of a court. Mark lives nearby and sets up the Trangleball court every morning with the permission of the community. A player throws the ball to a teammate and, within three throws, one of them has to aim for a face on the pyramid to score a point. A game is complete when one team reaches eleven points; that is, like volleyball, the serving team scores a point when one of its players throws the ball into the pyramid and the opponents don't catch it.

So far, about 275 Trangleball kits have been sold to YMCAs, summer camps, even prisons. Recently, Mark has shipped some as far away as Australia and New Zealand.

Mark pushes Trangleball every chance he gets. "I'm the pied piper for it, and I'd like to recruit more pied pipers to make this game grow." At least one pied piper is

influencing an entire nation. "A few years ago, a recreation student from the Czech Republic did his entire thesis on Trangleball, he loved playing it so much on Fire Island. At first, his professors wouldn't accept it as a serious idea, but he was so persuasive, he earned his doctorate based on this thesis. In 2001, he invited me to his university, where I helped him introduce the game to recreational educators.

"I'm really a shy person. I had to make a speech in front of 1,000 people in the University's gym with the governor and the mayor present. I had stage fright. Now, I'm making speeches all the time to promote this game."

Working this angle full-time for the past four years, Mark is on fire about promoting Trangleball. "This game has given me a reason to live. Before, I was directionless. Now, there's a sense of destiny for me. The journey I've been on has made the whole thing worthwhile."

He's aiming in a clear direction to raise the profile of Trangleball and sell the licensing rights to it. "If it takes off, I know I'll always have security teaching and coaching it. I'd like to see a big company manufacture, market, and distribute it so more people can play."

Right now, Mark is having a ball promoting his game. But he knows he'll have scored big when Trangleball becomes a household name and a game played on every corner.

STAT BAR

PATENT: pending

PRODUCT PRICE: under $200 with shipping

STATE: Illinois

INVENTOR'S AGE: 51

INVENTOR'S PROFESSION: investor and entrepreneur

MONEY SPENT: $70,000

MONEY MADE: $30,000–$40,000 (and still has inventory to sell)

WEB ADDRESS: shootAndstar.com

" . . . this time I decided I wasn't going to wait for some large corporation to mess things up."

shootAndstar Rebounder™

Basketball for One

Talk to Glenn Hudson for just a few minutes and you'll learn he truly and completely loves the game of basketball. "I came from a family where my dad didn't want me to waste my time playing sports. Even though I started both my junior and senior years in high school, he didn't ever see me play until after the whole town was going crazy over our team's success."

For the small Illinois town called Gibson City, winning the state games back in 1972 was a very big deal. Now, years later, Glenn has taken that love of basketball and created an invention that may help train future b-ball stars.

Learning early on that shooting hoops over and over again was the secret to sinking those basketballs, Glenn has scored with the "shootAndstar Rebounder," which attaches to a standard backboard. It enables a lone basketball player to practice shooting easily. Shoot from any place on the court, and both made and missed shots are quickly returned to the shooter.

Glenn remembers, "Our team's star player, Dennis Graff, led the state in scoring and ended up getting a full-scale scholarship to the University of Illinois. One night before a regular season game, Dennis took me with him to a gym in a small town nearby, and we took turns rebounding for each other, rapidly passing the ball back to the one shooting for another shot. We did this for a couple of hours, and I remember that I was so confident about my shots before the game that I told one of my teachers, Jim Clemons, I was going to score a lot of points that night. I scored 26 points, the most ever, and the headlines were 'Graff Cold, Hudson Hot in Gibson Win.' Dennis scored 20 points but, as he pointed out to me, he still got the first headline."

Before coming up with shootAndstar Rebounder, Glenn developed and obtained a patent on a multisport backstop that let sports enthusiasts in golf, softball, soccer, football, and other sports practice their game alone. The device got picked up by a company called Sportime, Inc. "Each time I called the representative about the product, while they were preparing to manufacture it, he told me what a great product it was going to be. Eventually he told me that the overseas engineers couldn't overcome some engineering problems, and they weren't even going to manufacture the product. At the time, I didn't have the

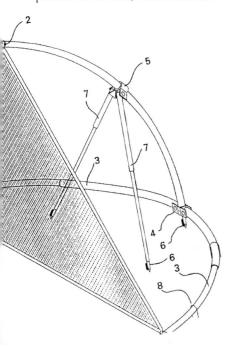

energy level or means to continue forward with the product."

Glenn was down, sure, but not out. You see, Glenn rebounds quickly. One morning, about 3 a.m., he woke up with the idea for a new invention. "With the experience under my belt of my prior invention, it was a lot easier to get through the prototype stage. I came up with improvements, and this time decided I wasn't going to wait for some large corporation to mess things up."

Since everyone, including his wife, Pat, had gotten so excited about his first invention, friends and family all took a "wait-and-see" attitude. At first, sales for shootAndstar Rebounder were terrible, so much so that Glenn thought of calling it a game. But he kept remembering a story he heard about the inventor of the shopping cart.

According to Glenn, when the shopping cart was first invented, no one wanted to use it. People have trouble making changes, and everyone was used to bringing their own baskets into the store. So the inventor paid people to push shopping carts around the store, taking things off shelves and putting them into the carts. Other people saw them and eventually started using shopping carts, too.

Anyway, over time, sales of shootAndstar Rebounder keep picking up. Glenn has now sold units to customers in almost

Perfect for a game of one-on-one.

forty states, a couple in Canada, and as far away as Australia. "I did make some mistakes with my initial inventory of shootAndstar Rebounders that I had manufactured. But none were so glaring that it wasn't still the best shooting practice device in the market for the price."

These days, Glenn is working on a new one-size-fits-all shootAndstar Rebounder with a company that manufactures products for such companies as Mattel and Fisher Price. "It will be a better product, with a lower cost, and won't take as much money to ship. After spending very little money in marketing and having initial success, I know that one day soon you will see the shootAndstar Rebounder in every neighborhood in this country and throughout the world!"

He shoots, he scores . . . and he takes over the world.

SwingSrite™

In the Swing of Things

STAT BAR

PATENT: US #6767059-B1

PRODUCT PRICE: $1,235.00

STATE: Illinois

INVENTOR'S AGE: 66

INVENTOR'S PROFESSION:
contractor; former plasterer

MONEY SPENT: $10,000

MONEY MADE: sold about 20 units

WEB ADDRESS: r-wayinc.com

"The challenge was to persuade the patent office that my design was unique."

Up and down and up again. Shirley Rhoades had always loved swinging in a swing. It was a lifelong joy. But when she got osteoporosis first in one hip, then the other, swinging became painful and then unbearable. It looked liked Shirley's days of swinging were over. She was down, but not out. Shirley's husband, Dorman, knew there had to be a way to get his wife swinging again.

Dorman Rhoades studied the problem and took a swing at the solution. He designed and built a wooden swing with an adjustable steel subplatform for Shirley's feet. On this platform is a treadle controlled by the feet, which serves to take pressure off a swinger's back. He used a special airplane cable and made adjustable seat slats out of white oak. By the time Dorman was finished, his swing was so unique he could tout fourteen patent claims on it.

With this design, the swinger's feet generate the swinging movement. "The treadle is so sensitive, it can be moved with slight movements of the toes. The platform has a rolling action in the metal undercarriage under the wood. It changes the center of gravity and made it possible for Shirley to relax and swing for hours, even before she had both her hips replaced."

This fifth-generation plasterer, who's become a general contractor, started designing and building this swing about six years ago. He called it "SwingSrite" and it took about four years to get the patent issued. "You see, swings have been around for hundreds of years. The challenge was to persuade the patent office that my design was unique. I thank my patent attorney for achieving that."

Taking the swing to new heights

Beyond placing SwingSrite in their backyard, it makes an attractive addition to a garden or public place. "That's where I've had the most success in selling it. At Eureka College here in Eureka, Illinois—it's the college that Ronald Reagan attended—the senior class had a budget to do something nice for the college. They looked at my swing and bought two sets."

Dorman sees this invention as more than a relaxing swing and piece of landscape art. After selling one to a therapist for use with an autistic man, he thinks the market could be lucrative if he pursued buyers with disabilities. "The therapist loved the swing's design and stability. She trusted it enough that she feels safe leaving her autistic patient alone to swing for hours. The swinging motion calms him down."

Ideally, he could sell licensing rights to a swing company that would make a plastic mold and produce it for less than its current price of $1,235. "I've invested about $10,000, including $6,000 to get the patent. I've made and sold about twenty so far, so I'm probably in the black. At least I've recovered my costs." Because it's handmade with all stainless steel hardware, the SwingSrite is expensive to produce. And Dorman doesn't have a factory set up. It takes him at least two days with a helper's assistance to build each one specially ordered.

As a 66-year-old contractor who already works forty to fifty hours a week, Dorman doesn't have the time or interest to delve into marketing SwingSrite. But he sure enjoys inventing. "My mind is cranking all the time." He's invented a hammock that's easy to get in and out of. He's devised a method to make it easy for Shirley to move between the pier and their pontoon boat. He's even come up with a better way to barbecue burgers, using metal plates that go on top of and underneath a hamburger patty. You see, the plates have metal spikes that go through the patty to cook the meat evenly in the same way skewers in potatoes help cook them faster. And because of the holes from the spikes, condiments easily ooze into the heart of the meat.

Dorman is especially proud of an exercise machine he's invented. It's a three-wheeled side-by-side recumbent bicycle that lets Dorman and Shirley cycle down the streets of Eureka beside each other. The way he's put together the gearing, they can even be pedaling at different speeds just by adjusting the gears. "Shirley and I love to go pedaling down the road, talking and fully sharing the experience."

I'm sure you've realized what is obvious to me now—that the story of Dorman Rhoades's inventions is really a love story. Anyone have a handkerchief?

Pohol™

Fishing Hands Free: It's No Fish Tale

STAT BAR

PATENT: information not provided

PRODUCT PRICE: $16.95

STATE: Virginia

INVENTOR'S AGE: 44

INVENTOR'S PROFESSION: financial analyst/entrepreneur

MONEY SPENT: $26,600

MONEY MADE: nothing yet

WEB ADDRESS: pohol.com

"...you want to be like a rat in a maze: If someone blocks one hole, you've got to find another hole to go into."

Anyone who knows how to fish will tell you it takes more than two hands to do the deed. To catch a fish, you have to lay your rod on the ground, squeeze it between your legs, or pinch it under your arm to release the fish or bait the hook. Such precarious positions can damage a fisherman's precious pole.

After countless times of wishing he had a way to free up his hands—not to mention a way to stop damaging his fishing rods—Mareo Williams thought of an invention that would let him have his fish and grab it, too. He decided to put his pants to work and designed the Pohol to do the pole holding for him.

Mareo says, "The Pohol will make even the most inexperienced fisherman feel like an expert." Now how do we break the news to the fish?

The Pohol attaches to a fisherman's belt and creates a sort of holster for the fishing pole. This way, a fisherman has a place to rest his pole and his hands are free to grab the fish, bait the hook, or answer his cell phone.

The Pohol holster evolved in Mareo's garage. He and his partner, Miguel Maroles, made three prototypes before they hooked the design that worked.

Everybody loved it, except for Mareo's older brother. "He laughed at my idea. He's married to the city attorney, so he's set in his own ways, but his disapproval gave me the motivation to make it work."

They originally named the pole-holding device "The Poho." They got the name by abbreviating "pole holder" to "Poho." But Poho sounded like something else and Mareo didn't want to risk offending women. So no matter how many people liked the original name, the experts at the Yankee Invention Expo recommended a name change to Pohol.

Mareo drove to the patent office five times so he could hand-deliver his paperwork to the patent examiner. He explains, "That way, I got to know her, so if we had a problem, we could work it out. Since the drawing didn't tell the whole story, I took the prototype up to her. It made sense to her once she had it in her hand. After that, she put the patent through."

Mareo didn't start with such a direct relationship with the patent office. He went to lawyers first, but they wanted $5,000 to $10,000 to do the filing. By filing the patent himself, it ended up costing him only $600. All in all, he's dangled $26,600 on the end of the Pohol: $13,000 to make the tool, $600 for the patent, $7,000 on a TV commercial, and $6,000 in inventory that's sitting in his garage ready to go fishing.

The fish are in trouble now.

Since getting the patent, the partners have been fishing for customers. The Pohol was on *The Tonight Show with Jay Leno.* Mareo and Miguel ran a commercial in Minneapolis until it got too cold for fishing. They now plan to run the commercial on ESPN and the Fishing Channel. The bait's working; Mareo just might have a solid bite. One company is interested in licensing the Pohol and getting it into stores. This is a deal Mareo is interested in reeling in. "This company doesn't charge anything up front, like some companies do. It just asks for some of the royalties."

Mareo knows when to fish and when to cut bait. "I learned in college that you want to be like a rat in a maze: If someone blocks one hole, you've got to find another hole to go into. You don't want to be in a position where someone can squeeze you. So I keep adding trades to my list of professions." At last count, his list included realtor, beauty salon owner, financial analyst, entrepreneur, stockbroker—and now inventor.

Mareo is reaching for success, and with the Pohol, his hands are free to grab it.

STAT BAR

*"Without the Scope Mate,
it's just trial and error."*

Scope Mates®

No Turkey Here

Larry Muncy of Lake Milton, Ohio, is a turkey shoot enthusiast. Animal lovers needn't worry. Participants only shoot at targets. The one closest to the "X" wins. To make the game really exciting, there's money at stake. Surprised?

Anyway, Larry is a good shot and he loves turkey shoots. "I get there early, take a couple of practice shots, and sight the scope back in. We only shoot six rounds. There's no time to adjust the scope.

"I shoot from different distances: 20-, 22-, 25-, 28-, and 33-yard shoots. If you're off an inch or so at 100 yards, you take four clicks to readjust your scope, so one click moves it ¼ inch. At fifty yards, one click is ⅛ inch. At twenty-five yards, to move it one inch is sixteen clicks. But I don't have time to do the math during the shoot."

One day, Larry sat down and figured it all out. The ¼-inch scope is the most popular, since it's the least expensive, so he based his measurements on that. He took a 6-inch ruler and marked every ¼ inch on a piece of paper. Then he did the same thing for various distances. Bull's-eye! With support from his shooting buddy George Pedigo, Larry came up with a quick and precise way to tell how many clicks were needed to zero the gun scope to target center, for any distance and type of scope. He named his invention Scope Mate.

The next step was zeroing in on a prototype. "I found a guy in Middletown, Ohio, who made 6-inch to 36-inch rulers out of plastic. I showed him my idea, and he said he had some extra plastic he wouldn't charge me for and he'd make as many scopes as he could from it. It only cost me $380 for twelve of each gauge." One Friday, Larry and his brother-in-law, Michael Roth, went to a big money shoot. "I shot it, and shot left. He shot it, and shot left. Then

I put my Scope Mate on. It told me to adjust fourteen clicks. I did and won $375. It done paid for itself."

Larry's wife, Sheila, used to call him a dreamer because of his get-rich-quick schemes. But one night, after working the late shift, he was watching TV. Someone on the show said, there's only one way to get rich, and that's quick. You're either born into it, win the lottery, or invent something. That last one hit hard. He had missed out once. "I opened *Popular Mechanics* one day and there was a wrench I designed—but didn't pursue—selling for $49. I was sick! So when I invented the Scope Mate, I told my wife about it and she got me the number for an invention submission corporation."

Larry spent $900 on the patent search. The corporation wanted $3,500 for the next step, so Larry called a patent attorney. In the meantime, he ordered a Provisional Patent Application (PPA) kit on the Internet. He and his sister, Drema Roth, completed the application and turned it in. It was so good that the patent attorney recommended a patent agent, who converted the PPA into a patent for only $2,500.

Larry's original prototypes were made from clear Plexiglas. They were beautiful, but they scratched too easily. Larry found a manufacturing company that made 1,000 sets for $800. He also found an interested retailer. "If I was to package

them and give him a corporate letter, a local sporting goods store owner said he would put them in his store. But I can't meet the demands for production. That's $30,000 I don't have." As a vet living on disability, Larry doesn't have the resources to mass-market the Scope Mate. So he's set his sights on licensing it, and would even sell the patent.

Larry knows the Scope Mate works—it's been field-tested and hits the mark. He says, "The Scope Mate's not for the gun, it's for the scope. If you've got a scope on your crossbow, pistol, or rifle, it'll work. If you can't hit the bull's-eye at 100 yards, you definitely won't at 200 or more. The farther away you are, the more off-target you'll be. Without the Scope Mate, it's just trial and error. You might go through a box of shells or more trying to hit that target."

One day, Larry passed a Scope Mate prototype to Pete, a turkey shoot friend, to try out at a shoot. Larry missed it, but Pete had gone. At the following shoot, Pete told Larry, "Man, I've been looking for you! I just wanted to see how full of bull crap you were, so I tried the darn thing." He'd been using his son's .22 rifle. Pete said, "I'd been trying to sight this thing in and couldn't do it. Every shot was off! So I put that gauge up there. It said move eighteen clicks, and you know, that next shot went right through the center. That thing works!"

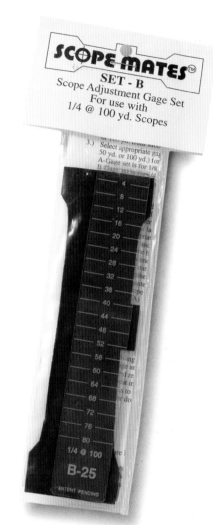

While aiming for success, this invention hits the mark.

Yep, Larry has set his sights on success and has a bull's-eye of a product. So far, production, distribution, and marketing has missed the mark. Anyone interested in scoping this one out?

Boomwhackers®

Wrap Music

"…if I can tune them, I can make music."

An empty roll of wrapping paper and a scissor don't sound like the raw materials for a successful invention. But those two items put Craig Ramsell into the world of invention and wrap music.

Craig was wrapping a friend's gift. As he used up the last bit of wrapping paper, Craig was left with the cardboard gift-wrap tube. He then mindlessly took the tube to his recycling bin for that week's pickup. "I freaked out. The tube was longer than the 2-foot by 2-foot recycling maximum. So I got scissors to cut the roll and wound up with one in each hand. I had no thought about playing them to make different sounds. I wasn't even a drummer. But they were whispering in my ear."

Craig heard the sound of music—Cha-ching!

"As soon as I hit the rolls on my thighs, I could tell they were different pitches. That's when I noticed that one was longer than the other. And I reasoned, 'If they're different pitches, I can tune them; and if can tune them, I can make music." If this were a movie and not a book, I would put in dramatic music right here.

Craig had been home on medical disability for radiation therapy to his head for a pituitary tumor. He'd been moping around, depressed about his medical situation, and while home he became friends with Pat, a retired nurse across the street. After banging the cardboard tubes against his thighs and the kitchen counter, Craig went across the street and rang Pat's doorbell. She opened it to see Craig hitting tubes on his head. "'Hey Pat,' I said, 'I've got this idea—.'

"She said, 'We were worried about you before. Now I'm really worried.'

"'No, Pat, I'm on to something here,' I told

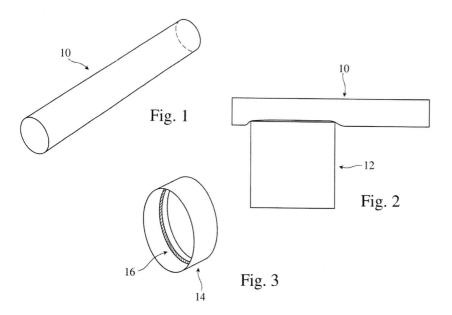

Fig. 1

Fig. 2

Fig. 3

good." Craig asked the store clerk about the material. Once he learned everything the clerk knew about plastic mailing tubes, Craig whipped out his "plastic," bought a couple of tubes, and started experimenting with different lengths.

Another serendipitous moment was about to whack Craig upside the head. Sitting around the house waiting for a guy to come fix his furnace, Craig leafed through the entertainment guide in the newspaper. Musician Zakir Hussain, from India, and his group would be kicking off a percussion festival.

"Tired of waiting for the furnace guy, I left to run errands around town and got hungry. I decided to try an Indian

The sound of success

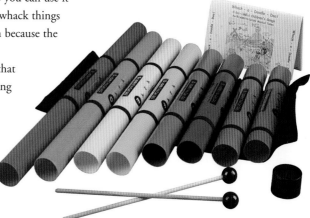

her. I knew from the first moment. I could have just as easily folded the tube to make it shorter. But I got the scissors. So many people, including Mom, said, 'Oh yeah, you and your brother used to bang paper towel tubes.' I took it to the next step—it was waiting for me."

Several months went by before Craig set foot in a plastics store. The cardboard tubes weren't holding up. He needed tubes made of flexible plastic that would behave like the cardboard, but last longer. He found a plastics store in San Rafael, California, where he lived at the time, and stopped in one day while running errands to see what they had. He found lots of tubes that were too stiff.

"The wall of the tube needs to flex easily for my invention to work. You can make sounds by hitting the end of the tube. But what makes my idea unique is that the wall does flex, so you can use it for body percussion and whack things without destroying, them because the plastic gives."

He found something that looked promising—mailing tubes. "I looked around to make sure no one was watching, took the cap off one tube, squeezed it, hit it—boop, boop, boop. It sounded

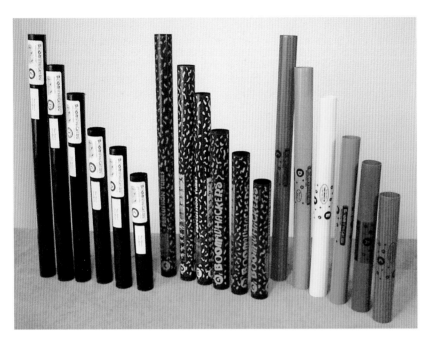

restaurant named Chinar. I was still moping, upset about the furnace guy not showing up, when in walks Zakir Hussain. He was wearing a Planet Drum T-shirt. *Planet Drum,* a book by Mickey Hart, Fredric Lieberman, and D.A. Sonneborn, is a book I'd decided I had to read. I needed to get smart about drumming." As if that wasn't enough for one day, Zakir was with a guy named Shivamani. Craig relates, "Monnie is my wife's name, and we named our dog Shiva. The other two-thirds of my family was in this guy's name. The synchronicity of it!"

Craig approached them and offered to buy them lunch. Zakir declined, since he was buying lunch for his friend, but that was enough to break the ice. Craig told him about his musical tubes, and how he could play them on his body. "Zakir said, 'Show up at rehearsal Thursday night and we'll try them out. If we like them, we'll do a world debut with them at the concert Friday night.'"

Craig didn't have any tubes tuned at that point. He only had stock in his garage and forty-eight hours to transform them into instruments. His MIT-molded mind worked the physics.

"It has to do with the way air travels out of the tube, like the tuning of wind instruments." He borrowed a saw from his neighbors and called around for a patent attorney. He was assured he had time to apply for a patent, even after the first public use.

He put the prototypes together and labeled them Pipe Drums. "At rehearsal, Zakir had us sit there for a couple of hours. He and his musicians talked about how they imagined the tubes had evolved over time, from rock to metal, and eventually to plastic. 'These are really cool; we'll definitely do something with these.'" Craig was in.

"The next night, we were frantically labeling tubes backstage, when an older gentleman wandered by, said they reminded him of something that might be used by some obscure island tribe, and asked about them. Then the same guy walked onstage to introduce the concert.

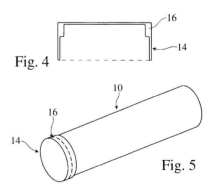

Banging tubes to make music

It was Fred Lieberman. And he wove my invention into the introduction, saying how people have beat things for thousands of years, and, in a while you will see that people even use plastic tubes."

When it was time to play the tubes, Zakir and his group came down the aisles, hitting themselves with them. People got a kick out of the head thing. Next, the musicians thumped each other, then members of the audience. Thrilled, Craig got it all on video. He couldn't believe his luck. He was pinching himself. One month to the day from when he set foot in the plastics store, a world-class musician was performing with Craig's invention in public, onstage, in San Francisco.

A couple of days after the concert, Craig found his inspiration to change the name because of a write-up in the newspaper. The opening line was,

"People have been whacking things for thousands of years." Whack. That resonated. He combined different words with it, playing around until he had fifty words scribbled down. "'Boomwhackers' jumped out at me. I liked the sound and flow of it. That was when the word 'whack' began to thump into my consciousness.

"I always thought of myself as the reluctant entrepreneur," Craig says. But he didn't see himself running a business and tried to find a company that would license it. "Because of my personal situation, I tried to create a minimum monthly income. Nothing came of that, but individuals called to make orders." He was selling his Boomwhackers one whack at a time.

Then, at a convention in Washington, his sales started booming. They had brought in Arthur Hull, a fan of Craig's Boomwhackers, to do clinics. The participants streamed straight out of his clinic to the sales table. "Off we went. It was a major jumpstart into the education market." In an era of low-budget music in schools, the

Boomwhackers fill a niche as an inexpensive, useful, music educational tool. Besides, the fact that they're so engaging, so whackable, is a plus.

"We're transforming music education in countries around the world. About half of the units we presently ship go to over two dozen countries." More than 4 million units have been sold through Craig's company Whacky Music, Inc. and its retailers and wholesalers.

As Craig likes to say: "There's no such thing as out of whack. Something can be whacked out, but it can't be out of whack." His invention sure isn't, and business is booming.

STAT BAR

PATENT: information not provided

PRODUCT PRICE: $69

STATE: New York

INVENTOR'S AGE: 35

INVENTOR'S PROFESSION: engineer

MONEY SPENT: $600

MONEY MADE: $0

WEB ADDRESS: pulseclock.com
(domain name reserved, not up yet)

"It's easy to invent something. The hard part is finding a problem that needs a solution."

Pulse Clock™

And the Beat Goes On

I would assume most inventive ideas start in the brain. But Victor Konshin's invention brainstorm started in his heart. You see, Victor got his invention idea while taking his own pulse.

"I was at the gym, working out on the cross-trainer machine, when I wanted to check for my target heart rate. The built-in pulse monitor on the machine said my pulse was 299 beats per minute—which, if accurate, would mean I was about to drop dead!" He didn't. So Victor came to the conclusion that the pulse monitor was broken. "I decided to measure my pulse using the method described on a poster on the wall: Count your heartbeats for ten seconds and then compare it with the chart. It's difficult to count seconds and beats at the same time. Our brain isn't set up to count two different things at once. I wondered if there was a better way."

It's that wondering that makes Victor's inventor mind tick. "I've always had a creative approach to problem solving. It's easy to invent something. The hard part is finding a problem that needs a solution. Most people accept things the way they are, and don't see a problem. But an inventor thinks, 'How can this be done better?' I've had millions of ideas for products, and I'm getting sick of seeing them come to market with someone else's name on them—and the money going into their bank accounts."

Because he couldn't turn back the clock, Victor got busy and came up with an idea for a new clock. His first impulse was to design a box to hang on the wall with a face that would light up every six seconds. The light would signal to those exercising that this was the time to count their heartbeats. They'd take that number and multiply by ten (or add a zero) to get their beats per minute.

But Victor decided a box on the wall was cheesy. Besides, people are always watching the clock. So he bought a cheap clock at Kmart, and adhered a circuit board, with lights and the timer to the back. The Pulse Clock has a ring around the edge that lights up red (with the most powerful lights he could find) every six seconds.

An engineer for a computer company and an electronics hobbyist, it only took him two tries to get the Pulse Clock cooking. "The first one didn't work too well. I got something wrong in the circuit board, so I scrapped it." The second one worked like a charm.

"Using an existing clock would reduce manufacturing costs. I decided on the timing—six seconds on and four seconds off—to allow for two different ways to count. Those without the benefit of a conversion chart can count their beats for six seconds (while the light is on), and then multiply by ten to get their pulse. If they do have a 10-second conversion chart (or are good at math), they can count from the time the light goes on to the time the light goes on again

(six + four seconds) to give them an accurate reading."

Victor is still debating whether to go to market with the Pulse Clock. He's been working around the clock on another idea for which he recently got funding. "Right now, the Pulse Clock is in the prototype stage. I'm in the process of starting up a company to produce this and other product ideas. The

Lights stay on for 6 seconds—count your heartbeat while lights are on, multiply by 10, and you have your heart rate.

company will be an innovations company, marketing both my ideas and those of other independent inventors."

Victor's efforts aren't in vain. The Pulse Clock can't be beat, and Victor knows it in his heart.

HanuClaus Hat™

Merry Hanukkah, Happy Christmas

STAT BAR

PATENT: US #D501980-S

PRODUCT PRICE: $14.99

STATE: New Jersey

INVENTOR'S AGE: 41

INVENTOR'S PROFESSION:
graphic artist

MONEY SPENT: approx. $4,000

MONEY MADE: no profits yet ("I just
started manufacturing . . . but, I did
sell a few of the hats I made personally . . . and made approximately
$400")

WEB ADDRESS: hanuclaus.com

*"I lucked out on being able to make my
own prototype and not having to hire
a manufacturer to make a mold—that
saved a lot of money."*

Sometimes when you least expect it, you hit on what could
be a great idea. That's what happened to New Jersey resident
Cynthia Siebert.

Cynthia is a graphic artist who volunteered to be the party coordinator for her
office holiday party. She designed a holiday card to send to the other branch
offices. They took a group photo and
decorated the card with graphics. In
the photo, one of the two bosses
was dressed in a Santa costume
and the other was sitting on
"Santa's" lap. (This is how
Cynthia's bosses behave even
without eggnog.) Anyway,
Cynthia thought it would
be cute if she inserted a
graphic of a Santa hat
on the boss who was
sitting on Santa's lap. It
was cute. Then Cynthia
had the idea of adding a
dreidel on the end of the hat. A dreidel
is a four-sided top with a Hebrew letter
on each side and it's one of the best-
known symbols of Hanukkah. Because

both of Cynthia's bosses are Jewish, it
made sense.

Her co-workers were amused by her
design and encouraged her to bring it to
life. Cynthia felt if she could
make that happen, it would
be a Christmas miracle,
or maybe a Hanukkah
miracle. With the
encouragement of her
friends and
co-workers, Cynthia
jumped into the hat busi-
ness headfirst. She started
the patent process in
2000 and, five years
later, in March 2005, she
received her patent on the
HanuClaus Hat.

"There are so many families who
celebrate both Christmas and Hanukkah.
One parent might be Christian and
the other Jewish. And many people are

of mixed ethnic and religious backgrounds. Why not share the holiday season by bringing people closer and celebrating both traditions together?" Cynthia reasoned.

At first, she didn't want to go public with her idea because she was afraid someone would steal it. But since receiving the patent, she feels protected and no longer worries about it. She was able to keep the costs down by making her own prototype and having the hats embroidered at a T-shirt place. "I lucked out. My patent was only $1,500 because it's a design. And I lucked out on being able to make my own prototype and not having to hire a manufacturer to make a mold—that saved a lot of money." The few she made went over big. Cynthia sold fifty just by word of mouth to co-workers and friends, and then from her website, hanuclaus.com.

Cynthia took her prototypes to the Yankee Invention Expo to get a feel for how the public might react to the HanuClaus Hat. After receiving a lot of positive feedback, she decided to move forward into full-fledged production. She just needs to find the right Santa's workshop and she's currently getting pricing on mass production.

Christmas came early in the fall of 2005 when The Tonight Show with Jay Leno gave her a call. Cynthia and six others were picked from eighty hopefuls at the convention to be included in his skit called "Pitch to America." Cynthia also got to present her product on a few local newscasts. What really got her going, though, was the introduction of "Christmukkah" on Fox's TV show called *The OC*. Cynthia figured if this holiday combo is going mainstream, she'd better move on it. A few weeks later, Cynthia got the ultimate omen that her HanuClaus hats were meant to be. That year, Christmas and Hanukkah fell on the same day. That was it. Cynthia took that coincidence as a sign not to be ignored. "Someone was telling me it's time."

A lot of people celebrate both holidays and there's nothing else in the market to bridge both traditions. Cynthia has found a way to bring both cultures together. She's found a cozy niche, developing an entire HanuClaus line, including decorations,

ornaments, and items for pets. She hopes to get her products into stores in time for the coming holiday celebrations. "I can't wait to see the HanuClaus line on the store shelf."

Cynthia hopes the season of giving is very generous to her. Ho Ho Ho. Oy Oy Oy.

Hats off to uniting Christmas and Hanukkah

Harness Play Pack™ / Picnic Party Cloths™

What's in a Name?

"I was always designing things, but never dreamed of getting a patent."

Our next inventor is Nancy Kerrigan. No, not the Olympic ice skating star, but our Nancy Kerrigan also had a bad experience to tell about.

A man who our Nancy thought was helping her with an invention instead stole the idea and sold it to someone else. With the help of a lawyer and some money, our Nancy eventually got the rights to her invention back, learning through this ordeal the business behind inventions from the ground up. Speaking of inventions, let's talk about our Nancy's ideas: the Harness Play Pack—backpacks for dogs—and Picnic Party Cloths—tablecloths with a purpose.

The idea for the Harness Play Pack came to Nancy as she watched a parade with her husband, Peter, in Wakefield, Massachusetts. A woman who was walking in the parade with a German shepherd had attached a water bottle to the dog's collar. Nancy thought, "Wouldn't it be great to expand the collar idea into a full backpack, to carry even more?"

Happily, Nancy had the sewing skills to create it. She'd been sewing for twenty-five years. "My grandmother taught me how to sew." Nancy sowed her inventing seeds helping the Blessed Sacrament Color Guard in Cambridge, Massachusetts, while her daughter, Cheryl, won awards twirling a rifle in the color guard during half-time shows. Nancy made costumes, props for shows, curtains, and bedding. While working as a nanny, she sewed quilts on the side. "I was always

Why should your dog get away without carrying his fair share?

designing things, but never dreamed of getting a patent." That all changed with the Harness Play Pack.

The Harness Play Pack attaches to the top of an adjustable harness with a belly strap and comes fully equipped for a fun day at the park with your pooch. Inside the pack, there's a flying disk that doubles as a water dish, a ball, and a water bottle. Even pick-a-poop bags are tucked in.

Within hours of inventing the Harness Play Pack, Nancy came up with the picnic tablecloth idea. She made it fitted so the cloth wouldn't blow away, and outfitted it with pockets to hold utensils, paper plates, crayons, or a deck of cards. Nancy's mom was proud of her.

"When I showed these products to my mom, she said, 'Nan, God's given you a special gift. You have two special products here. You work on them.'" She now has a line of five party cloths: picnic, tailgate, craft, game, and scrapbook.

Nancy's dealings with the man who stole her ideas happened at about the time Nancy's mother passed away. "My mother was my best friend, so after she died, I was devastated." A page on Oprah's website on a day in 2002 was Nancy's inspiration to keep going. It read in part: "Once you have tuned out the noise of your life . . . you face the biggest challenge of all: to find the courage to seek out your big dream." Nancy says, "That's what kept me going after my mother died. That's why I keep going now.

"I've mortgaged my house three times. I'm in debt, but I have my own company. Over the last five years, I've spent $300,000 on this company. I had an idea that was so good I couldn't sit on it. This is what I always wanted to do, but I never knew it until I invented."

When Nancy decided to become an inventor, she called 411 and found out about the Inventors' Association of New England, which meets at MIT once a month. Now its secretary, she has done prototyping for the inventors there, and has

also joined other associations and inventors' clubs since. But it's through her role as secretary for the IANE that she got on the *Jane Pauley Show.* She's been on *The Tonight Show with Jay Leno,* CNN, and the Home Shopping Network, among others, and has gotten press in newspapers and magazines. Several universities have invited her to speak on inventing and bringing a product to market. "If someone comes to me with a problem, I can solve it. If I can't put it in the sewing machine, I have connections with other craftspeople who can create a prototype."

Nancy is now looking to license her products. "All I want to do is design. I have so many ideas in my head I can't get them out fast enough." You might hear Nancy's story and say she's skating on thin ice. But not our Nancy; she's a medal winner who has the confidence to go for invention gold.

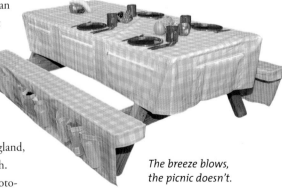

The breeze blows, the picnic doesn't.

Panic Mouse®

Cat & Mouse Game

STAT BAR

PATENT: US #6892675-B1

PRODUCT PRICE: $29.95
(mass-marketed version will be
half the price)

STATE: California

INVENTOR'S AGE: 41

INVENTOR'S PROFESSION:
full time at Panic Mouse Inc.

MONEY SPENT: $1,000,000
(made a lot of mistakes)

MONEY MADE: $1 million per year
for the past three years (much of
profits reinvested)

WEB ADDRESS: panicmouseinc.com

"We haven't looked back since."

There are 80 million cats in the U.S. That's a lot of cats.

I've been told that the difference between dogs and cats can be explained this way. When a dog owner provides a home, food, and affection for a dog, the dog thinks, "Wow, these humans do all this for me. They are like gods." But when a cat owner provides a home, food, and affection for a cat, the cat thinks, "Wow, these humans do all this for me. I must be a god."

Paul Comerford doesn't have a cat—he's terribly allergic to them. But he has several friends with cats. "They always have cheap cat products scattered around their houses that cats don't play with." Instead, the bored felines would simply scratch up the furniture.

Paul realized that the only time the cats would play was when people interacted with them, teasing them with a string of some kind. That's when, like a cat jumping onto a toy mouse, Paul got an idea. How about a robotic toy to replace the human? "The toy would need to have random and unpredictable movement, just like human arm movement, to entice cats to play."

So this non-cat owner went to Petco and Petsmart looking for a toy similar to what he envisioned. He even checked various cat-alogs. The only item even close was a stationary string that attached to a scratching post or door. All the other toys called for human interaction—and were priced under $10.

Having been in the import-export business, Paul contacted a friend in China and asked him to check factories there. He learned that his invention would cost around $16 wholesale and $30 retail. That's not birdfeed or even kibble.

After securing a patent, Paul contacted every U.S. distributor to see if any of them would be interested in working with him. No one called back. And when Paul talked to retailers, they told him no consumer would buy a cat toy priced over $19.95. That cat-astrophic news gave him paws for thought, but Paul still went forward.

After considering names, Paul decided on Panic Mouse, suggested by his friend, Eric Wu, in China. Next, he had a prototype made to show cat owners and people in the industry. The reaction was the same. "I found a real disconnect here between cat lovers and retailers." Pet industry professionals rejected the price while cat owners loved the idea and said they'd buy one at $29.95.

So Paul took the prototype to the Pet Expo in San Diego, open to both the industry and the public. Paul got over 300 orders during the two-day show. This was a problem, because Paul only had his prototype. He took the orders (no money) and said he would process their credit card orders once he had his cat toy in production.

Paul was now cat-egorically convinced of a market for his Panic Mouse at a price above $20. He joined the American Pet Product Manufacturers Association to display his invention at its trade show and Panic Mouse won Best New Cat Product at the show. The award brought

attention and Panic Mouse got a consignment test run at Petco.

Sounds like Paul's invention was easily winning this cat-and-mouse game, huh? Not so fast. Just as Panic Mouse was reaching for the cheese in 2002, there was a major port strike. Production got held up. Paul missed many deliveries since the ships weren't allowed to unload. He not only lost $350,000 in sales, but also got stuck with a warehouse filled with robotic mice. Now Paul was in a panic. "However, even with the delay in delivery, which could have killed our momentum, we had success in Petco that year."

Word of Panic Mouse's Petco success spread fast, and soon Paul's mice were in Petsmart, and on QVC and HSN. Then they were approached by Discovery Animal Planet to do a special production unit called the Discovery Animal Planet Hyper Mouse. It was a motion-activated unit that was among their top 10 percent best sellers in the store. "We haven't looked back since. We've sold 500,000 units without advertising and we're ready to launch our mass-market version under a different name brand in 2006."

Paul will be the first to tell you that he didn't do it alone. His wife, April, has been at his side doing the accounting and

This mouse keeps your cat jumping.

invoicing while raising their daughter, Madison. Paul also gives a great deal of credit to Eric, his partner in China.

So has Paul climbed to the top of the kitty toy heap? No way. He points out, "With 34.7 million cat owners in the U.S., we have a long way to go." Who would have guessed that a guy who is so allergic to cats would come up with the purr-fect product to cat-apult himself to success?

Me-ow, those puns hurt.

Clocky
BrightFeet Lighted Slippers
BOB
Channel Surfer
"Illum"-Gobo Illuminated Car,
Truck and SUV Lights
Rat Zapper
Enotrab Optical Alert Stand
Backlit Paper Cutter / Hover Pad & Bench /
Sea Stones
World System Clock
The InfoCube
Core Sound Trees
Peeping Thomas
Compact Disc Eraser

Electronics

I think it is safe to say that for most of us, our relationship with technology is a love-hate affair. We are obsessed with all things tech, but we also fear what we don't understand. And when technology doesn't work as promised, our frustration turns to anger as we head down the path of unreadable manuals, tech support and, ultimately, tech disposal.

Most truly high tech inventions don't come from garage inventors. Tech toys such as HD-DVD players, cell phone MP3 combos, and LCD TV's come from the bowels of mega tech companies. Sony, GE, Panasonic, Motorola, JVC, Nokia and many others crank out most of the tech hardware we've all grown to love. Tech gadgets that start out as luxuries quickly become must-have necessities. Those mega tech companies are all about gadgets for home entertainment, home office, and wireless everything. But those same companies have missed some obvious niche market products. That's where our garage inventors come to the rescue. Sony and Samsung don't produce a Rat Zapper.™ JVC and GE don't make bedroom slippers that light up. Panasonic and Whirlpool don't have in their catalog of products an alarm clock that jumps off the nightstand. But this next batch of inventors has tackled these products and more. I guess tech might be a four-letter word for many of us, but not for these fearless garage inventors.

Clocky®
You Snooze, You Lose

STAT BAR

PATENT: pending

PRODUCT PRICE: not yet determined

STATE: Massachusetts

INVENTOR'S AGE: 25

INVENTOR'S PROFESSION:
graduate student at MIT

MONEY SPENT: minimal

MONEY MADE: $0

WEB ADDRESS: clocky.net

"I like to think of Clocky as a troublesome pet that you love anyway."

Have you ever hit the snooze button so many times that you weren't sure what time it was? Whatever the time, one thing was certain—you were late. You're not alone. Too many of us have a love affair with our snooze buttons.

When Gauri Nanda shared this alarming observation with others, she learned that her fellow grad students had the same problem. While earning her master of science degree from Massachusetts Institute of Technology's Media Lab, she once woke up two hours late because she and the snooze button got too friendly. Fellow students who put their clock on the other side of the room were no better off because they knew exactly where it was. They'd almost sleepwalk to turn it off, and then jump back in bed.

When her Industrial Design Intelligence professor assigned his class to create an invention, an alarm went off in Gauri's head. She set out to design a clock that would really wake her up.

Gauri decided she needed an alarm that would run away from her—and wheels were the ticket. After all, a clock that rolls off your nightstand when you press the snooze button, rolls onto the floor, and wheels away might coax you into an early morning game of hide-and-seek. When the alarm sounds again, it's game time—you have to wake up to find it and turn off the alarm. What better way to wake up than winning a game with Clocky?

Not every student in that class came out with a patent, but at twenty-five years old, this inventor

Hit this snooze button and Clocky rolls off your night table and makes you chase him.

did—and she's on a roll! After making four prototypes, Gauri is ready to send Clocky into mass production. She's selling it directly on the Internet (after filling orders for her list of pre-buyers!) and wanted to have it in stores in time for the Christmas season.

"I like to think of Clocky as a troublesome pet that you love anyway. I used to have kittens that would wake me up and although I was annoyed at first, I couldn't help but be amused by them, especially since they are so cute," Gauri explains.

As an MIT student, she had access to some sharp minds and support from mentors who help MIT entrepreneurs get their ideas into the marketplace. Add entrepreneurial parents to the mix, and this inventor clocked some serious support.

Gauri says, "People e-mail with their funny can't-get-out-of-bed stories. One man uses three alarm clocks and then asks his friends to hide them. Another potential Clocky user wrote: 'The snooze button is my drug of choice; the sound of the alarm ringing means nothing to me.'"

Gauri knows that it is time to break the snooze button addiction with a new wake-up call. She hopes Clocky will put traditional snooze buttons to bed.

STAT BAR

PATENT: pending

PRODUCT PRICE: $39.95

STATE: Georgia

INVENTOR'S AGE: 62

INVENTOR'S PROFESSION: former
house builder, now inventor

MONEY SPENT: "We have made a
sizable investment in pursuing and
working toward a patent, developing
a high quality, functional product
and building the BrightFeet Brand."

MONEY MADE: "Our investment in the
product is really starting to pay off."

WEB ADDRESS: brightfeetslippers.com

*"At first, he taped mini-flashlights
to regular slippers to see if the
concept worked."*

BrightFeet Lighted Slippers™
Headlights for the Feet

A knock on the head was what pulled the trigger for Doug Vick's shot at fame and fortune.

Doug got up in the middle of the night for a glass of water. He turned on the kitchen lights, drank his water, turned off the lights, and proceeded into his very dark bedroom. Being a considerate husband, Doug didn't turn on any bedroom lights. He tried to climb into bed without waking his wife, Barbra. He almost accomplished his mission—until he slammed his head into one of the posts of his four-post bed.

Needless to say, Barbra woke up, the lights came on, and Doug saw a dark dilemma that needed a bright solution.

Doug Vick is not new to the invention world. Even as a child, he always wondered about the men and women who gave us products like paper clips and bobby pins. The 62-year-old Georgian was always a creative problem solver. Doug started as a home builder. One of his first inventions was called

Filtermate.® You see, Doug thought many homes didn't sell because they had a musty smell. Filtermate was scented foam rubber that could be put on an air conditioning filter. When the air moved through the filter, the home smelled like whatever scent was on the Filtermate.

If you think this idea stinks, think again. Filtermate became a big hit. It was in 20,000 stores including Walmart and Target, and Doug sold millions of them. The sweet smell of success wafted over to a Connecticut company, which bought him out in 1996.

Doug had a no-compete contract so he was out of the Filtermate business, but couldn't shake the idea of making cents from scents. One day, while visiting his daughter's new home, Doug didn't like what he smelled. The combination of new paint and dampness was very unpleasant. That's when he came up with "Paintpourri." You add a packet of Paintpourri to any paint and you've turned any painted wall into a giant air freshener.

Next, Doug came up with Filter Breeze.™ You spray it on your A/C filter and you can add an ocean or tropical scent to your home.

One day, while driving near his home, Doug's car hit a skunk. That encounter turned out to be bad luck for the skunk and good luck for Doug. He saw how quickly the skunk's odor

The late-night trip to the bathroom just got a little easier.

got into the car. What if that same logic were used for good smells instead of skunk smells? Doug added a scent to windshield washer fluid and invented Wiper Breeze.™

That brings us to Doug's heady meeting with his bedpost. He knew a little bit of light would have prevented that knock to the head. So Doug used his head to come up with a solution—lighted slippers.

At first, he taped mini-flashlights to regular slippers to see if the concept worked. It did. Now the real process of invention began. Doug first made a prototype with a manual on-off switch. The next prototypes had a weight sensor so the light would go on only

when the slippers were being worn. To save battery power, he added a light sensor so the slippers could be worn in the daytime without turning the lights on. Doug went through dozens of prototypes until his lighted slippers were ready for market.

Doug decided to call his invention "BrightFeet" and they took their first steps in the fall of 2005.

Doug believes he has a winner. And even though he's done much of his invention work with scents, he knows that with BrightFeet he smells success, not de-feet.

BOB™
The Anti-TV

STAT BAR

PATENT: information not provided

PRODUCT PRICE: $89 to $99—
when mass-produced, about $60

STATE: Colorado

INVENTOR'S AGE: Brian, in 30s;
Tom, in 40s

INVENTOR'S PROFESSION:
Brian: business consultant;
Tom: business owner

MONEY SPENT: $650,000+

MONEY MADE: "ready to sell first run
of 800 units"

WEB ADDRESS:
hopscotchtechnology.com

*"Parents have never had a technical
solution to media monitoring before . . .
until BOB."*

How could you hate anything called Bob? But if you're a kid who loves to watch TV, you better watch out for BOB.

As kids watch more and more TV, parents are getting more and more concerned. Parents see their little ones as "TV-aholics," addicted to that flickering box. Children are passing on "real" playtime with siblings and friends in favor of "electronic" play in front of the tube. That's why Brian Baker and Tom Gallop came up with BOB.

The number of TV screens in homes has, on average, tripled from one to three in the past twenty years. Brian says, "Our research tells us 92 percent of parents in the U.S. set rules to limit screen-watching of any type. But 75 percent of them fail to stick with their own rules. The kids' whining wears them down." In response, Brian would like to see everyone adopt a new family member—an electronic monitor named BOB.

"Parents have never had a technical solution to media monitoring before . . . until BOB." BOB is a device that's slightly bigger than a remote control and sits next to a TV, computer, or video-game system. The power cord from one of these units plugs into the back of BOB and is locked in place. Once BOB gets plugged into an electrical outlet, it's used to monitor the amount of time a designated user spends on the attached unit. Each of the users (up to six) has a four-digit PIN that must be entered before the unit can be turned on. And once a preset viewing time expires, BOB shuts off the system and won't allow it to be turned on by that user for the remainder of the defined period. I can already hear the screaming.

The buzz about BOB is building—with publicity in *Popular Science, Colorado Magazine, Home & Garden, Wall Street Journal, Time* magazine, and on TV networks including ABC and CBS. The goal is to reduce a child's weekly "screen time" from thirty hours to ten hours. It could put more play into children's lives and

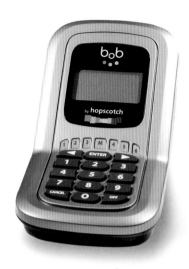

more harmony into their families, or it could turn some happy youngsters into very angry kids. Gulp.

How did BOB come into existence? Rewind to the spring of 2005 when Brian and Tom, both newcomers to Boulder, Colorado, met at a neighborhood picnic. Brian was living at a friend's house, recovering from "a near-marriage experience" in Manhattan. Tom, his wife, Michelle, and their three kids had just moved west to get out of New Jersey.

"Neither Tom nor I was looking for the 'next big thing' in our lives, but serendipity stepped in." Discussing the complaints of parents over too much media watching at this picnic, they put their heads together and envisioned an electronic referee with the name BOB. I'm sure you are wondering what BOB

stands for. The answer is nothing. They just picked BOB because it sounded like it would be a friendly guy for a family to have around. And after all, BOB spelled backwards is still BOB.

Together, Tom and Brian co-founded Hopscotch Technology™ to give BOB a home base. Brian, who's in his 30s, came from a management and consulting career in San Francisco, working with lots of start-up enterprises. Tom, who's in his 40s, still heads a family-owned dental manufacturing company, SS White, in Lakefield, New Jersey. He decided he could cut back on his hours at SS White to spend twenty hours a week as CEO of Hopscotch Technology. Brian, who works full-time and then some, is Hopscotch's president.

Before sinking their personal wealth and funds from angel investors into BOB's development, Brian and Tom did a series of research studies. The first one was fun and straightforward: they spent time hanging out at a local upscale mall and approaching parents to fill out a short survey in exchange for two bucks.

"There we were, two geeks in a mall, asking 100 people from a wide range of demographics to answer our two-minute written survey. We learned that all the parents—but especially those with an annual household income over $40,000—saw the need for a good way to manage media time for their kids."

This survey confirmed in a small way that they were onto something big. So they commissioned an independent market study that reached 4,200 people nationwide. "Statistically speaking, we wanted to go way overboard gathering data on this issue before we made the first prototype."

Once they had the assurance consumers would get friendly with BOB, they did a third test and put working prototypes into the hands of forty families. They talked to these families every day by phone to get a full sense of their experiences using BOB. Who participated in this study? Explains Brian, "We let this sample randomly select itself. We put an ad on Yahoo! classified asking for volunteers to test an electronic device. They'd receive $200 when they returned the test unit."

Sorry, kids, the TV plug is literally locked into BOB.

The results were very encouraging. The families (or at least the parents) didn't want to give BOB back. The parents instantly recognized how it could help them do their jobs as parents better. Many said they didn't even know their kids were watching TV in the middle of the night or that the time added up to more than thirty hours a week. "Having BOB around stopped the family arguments. 'Thank goodness' was a common comment from parents. I'm sure the feedback from the kids was not so glowing."

Using BOB shocked these families into waking up to how media was controlling their family life and throwing it out of balance. The results gave Hopscotch Technology the encouragement to make more units, go through the hoops of earning safety certifications, and get ready to launch BOB.

After investing $650,000 so far, Hopscotch Technology is working with Technology Driven Products® (TDP) in Loveland, Colorado, to build an initial run of 800 units. "We're marketing through our Internet site, and we've lined up three retailers. A total of fourteen retailers have asked to carry it when we're ready for higher volume, but we're going slowly at first.

"We have offices that overlook the Pearl Street outdoor mall in Boulder. From our second-story windows, we can watch the kids play on the sculptures designed for them. It's our reminder every day about the value of childhood play. I'm a true believer in our purpose." As stated on their website, their purpose is to "create products that reflect our company's mission: restoring Balance through Technology."

In Boulder and beyond, they've attracted the attention of lots of "believers" who really want BOB to succeed. More support from believers—plus more venture capital—will put BOB on guard watching over America's TV. "We know BOB has the potential to make a dramatic impact on our society, and really, to give kids back their childhoods."

At the beginning, Brian and Tom are setting a retail price of $89 or $99 for BOB. "And if enough believers become buyers, we'll see the price come down in no time."

But I think BOB had better be careful. I wouldn't want to be the one standing between TV-starved kids and their favorite TV shows. And kids do know where Dad stores his tools. BOB, you've been warned.

Channel Surfer™

Surfs Up, Dude

Who says a couch potato's brain goes to mush while sitting in front of the TV? For James Stokes, flipping channels triggered his creativity. "I was sitting back, pointing the remote at the TV, and changing channels, when I started thinking about how often I hear people say 'I'm channel surfing.' Wouldn't it be neat to make it real and have a surfboard-shaped remote control?" He went to work designing a cover for his remote control that looks like a surfboard. Dude, that idea is totally awesome.

Once you have a surfboard remote, you might want a stand for that remote. James thought a stand shaped like a wave was the only way to go. "With the stand, you don't lose the remote. It looks so cool sitting on the wave, you want to keep it there! That way, you're less likely to have to hunt for it in the couch cushion." James ought to know—his recliner has eaten a couple of remotes. "I'd sit back and hear a crunch—oops, it's time to go to the store again."

James sees limitless possibilities for the Channel Surfer. "Look at covers for cell phones—they come in different colors, shapes, even styles that light up. The Channel Surfer could come decorated with cartoon characters, or in sport themes with a helmet and the team's colors on the surfboard.

"I first told the guys at work, 'I've invented something, but can't tell you what it is.' From the time I filed for the provisional patent, I led them to believe it was a kitchen item. When I received the patent, I brought the drawings to work, and everybody thought it was so

STAT BAR

PATENT: US #6769658-B2
(Aug. 3, 2004)

PRODUCT PRICE: $25 for both cover and base

STATE: Florida

INVENTOR'S AGE: 39

INVENTOR'S PROFESSION: construction (builds roof trusses)

MONEY SPENT: $23,000 over five years

MONEY MADE: nothing yet

WEB ADDRESS: not yet

"They said I'd be an instant millionaire. But I know it's not that easy."

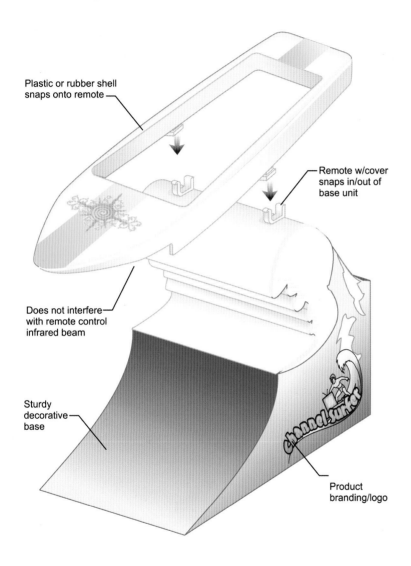

Plastic or rubber shell
snaps onto remote

Remote w/cover
snaps in/out of
base unit

Does not interfere
with remote control
infrared beam

Sturdy
decorative
base

Product
branding/logo

ChannelSurfer

cool, they were putting in their custom orders. They said I'd be an instant millionaire. But I know it's not that easy."

James has gotten full support from his family. His father lent him the money for the utility patent because he's sure it will sell. His dad just thinks that the idea needs to get to the right person in the right company.

James hit a few rough waves while trying to get his surfboard remote patented. The consulting firm he'd hired used a different attorney to pursue the utility patent. The patent office rejected it because the ink used in the drawing wasn't dark enough to make a copy. The same artwork was used to get the provisional patent, which was approved without a hitch. That didn't seem to matter to the consulting firm. They wanted more money to have an artist go over the drawing. Cha-ching!

After going through all of that, the Channel Surfer patent was rejected because the wording made it sound too similar to the patent description of some toys, and would interfere with certain existing patents. After a couple of rejections, the consulting firm said they couldn't do anything for James. In the meantime, his attorney had passed away, and his case had been sent to someone else. The consulting firm wanted more money to look at it again. Cha-ching!

Deciding he'd put out enough, James had the firm send all the info to him. "I got lucky. I connected with a helpful person in the patent office. I told him I was on my last string. The consulting firm was not helping. My attorney had passed away. And I do construction for a living—I don't know about legal writing. He told me how to rewrite it, what explanations I had to send to the examiner to prove that it wouldn't interfere with others' patents, and how to show my product was a cover for a TV remote control, not a remote control toy."

After filing the paperwork all by himself, James received his twenty-year utility patent. "I learned a lot at the Inventor's Expo, and wished I had done things differently. I paid over $6,000 to get the provisional patent, but it should have cost only a few hundred dollars. I paid the same company another $6,500 to get the utility patent."

The consulting firm had instructed James to send letters to presidents and CEOs of companies. It sounded like a good idea: The consultants found the contacts, and James just had to sign the letters, address the envelopes, provide the postage, and handle any ensuing correspondence. And if people showed interest, then he'd send them to the consulting firm and they'd handle the negotiations for 10 percent of the sale—

but the process never got that far. "At the Expo, I learned I could find contacts myself online for a lot less money and effort. And it's better to call the company's 800 number, find out who's in charge of its research and development department, and send it to that person, not to the president. Presidents don't open their own mail. Their assistants probably threw my letters out, thinking they were junk mail."

Still, James considers himself lucky. "I met people at the Expo who had spent $50,000 and had nothing to show for it. Consulting firms had told them to mortgage their houses in order to finance the product when they had no idea if their products would sell. Consulting firms will say anything is a great idea, because either way, they're getting paid."

After five years of work and $23,000 invested so far, James now has a patent for twenty years. Ideally, he'd like to crank up a partnership with a universal remote company. "I provide my patent; the company provides the money for marketing and production. It already has facilities set up for manufacturing and has relationships with retailers. The right company could

put it right on the shelf. I just need to save up a bit more money to have a professional prototype made."

So far, there's been no wipeout for this surfer. It seems like he's been mostly learning on the beach. He's still waiting for his big wave, so he can hang ten and shoot the curl.

You go, dude! Change those channels, cowabunga style.

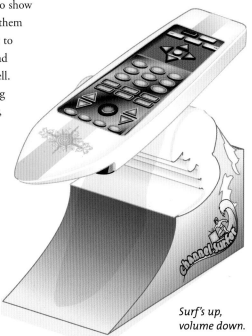

*Surf's up,
volume down.*

"Illum"™-Gobo Illuminated Car, Truck and SUV Lights

Lite Idea

STAT BAR

PATENT: #6685347

PRODUCT PRICE: N/A; retail price estimated at $100

STATE: born and raised in Texas, currently living on the road

INVENTOR'S AGE: 55

INVENTOR'S PROFESSION: Truck driver

MONEY SPENT: approximately $30,000–$40,000

MONEY MADE: none yet

WEB ADDRESS: marketlaunchers.com/grutze.html

"It's a testosterone thing. Everyone wants to have the better, badder truck."

Bright ideas hit inventors everywhere. For trucker Glen "Skip" Grutze, his epiphany occurred while driving down a lonely stretch of highway. "It was the middle of the night, and I noticed that all the cars on the road had headlights in the front and tail lights in the back, but nothing, no lights on the side. I thought, 'Something else is possible here. I don't know what, but there's a possibility for lights on the side of a car.'"

A Texan who now claims the road as his home, Skip drove on into the night, wondering how he could install lights onto the side of a car. And although the road is home, it took getting up into the air for Skip to find his next inspiration.

"On a flight to Mexico City, I was flipping through one of those catalogs they put in the back of the seat and landed on a page advertising projectors. They were inexpensive and you could position them on your driveway and shoot messages like 'Happy Halloween' onto your garage. That idea really grabbed me."

When Skip got back from his trip, he got to work. He researched the projector, called a "gobo," and started asking, "How could I make it better? How could I make it smaller?"

Skip's idea was to install a projector onto the sides of a vehicle to beam drivers' favorite emblems onto the ground as they drive at night. Car aficionados could project their car insignias; sports fans could project their favorite team logos. Skip decided to call his lights "Illum"-Gobo Illuminated Car, Truck and SUV Lights.

As someone who has been around the auto scene all of his life, this trucker knew the product's potential market was huge. Let's face it; lots of drivers are crazy about customizing their cars.

"It's a testosterone thing. Everyone wants to have the better, badder truck.

"Every Friday night of every week, people in every town and every city in this country are going and doing, hanging out. Kids are cruising and showing off their 'rides.' Whether it's a game or sporting event or just dinner and a movie, everyone drives and wants to be seen."

But soon this trucker came to a fork in the road. "Do I sit on this idea?" After all, Skip is a divorced dad with a teenage son. "Do I let it languish? Or do I get serious, step outside my comfort zone, do something bold, something wild?" "Something wild"—like become an inventor?

Skip shifted gears and chose the road less traveled. He enlisted the help of a friend to build a prototype, then hired a patent attorney. Two years later, he had his patent in hand.

But the patent was issued in 2001, around the time of the 9/11 terrorist attacks. Skip found himself at yet another crossroads.

"The world changed that day. I didn't know where the car market would be going, if the bottom would fall out, if people would be able to afford such products any more." Whether or not to

Projecting a logo onto the road

go forward with his invention was a serious decision and Skip "stayed up late thinking long and hard."

After a few weeks, it took some optimism and even patriotism for Skip to drive ahead with this. "We will survive . . . we'll lick the bad guys."

Several years later, Skip is still looking for a manufacturer for his product. He is encouraged by e-mails he gets from people expressing interest worldwide—everywhere from Hawaii to Dubai. But

without a manufacturer and distributor, Skip realizes that the vision he had years ago on the highway still remains a dream.

"I can see it, but at the same time, I won't believe it until a deal is signed on the dotted line. That's when I'll have to pinch myself. Until then it's still a dream, a dream of a better life for me and my son. But I do feel like it can happen."

While Skip Grutze continues to see high beams in his rear view mirror and taillights ahead of him, the sides of cars remain dark. Until Skip finds the right manufacturer, his "Illum" remains just a bright idea parked on the side of the road to success.

Rat Zapper™

A Shocking Rat's Tale

*"In the end, I think all the failures
made it better."*

In the 1800s, Ralph Waldo Emerson said, "If a man can write a better book, preach a better sermon, or make a better mousetrap than his neighbor, though he build his house in the woods, the world will make a beaten path to his door."

I don't know if that's true about a "better book," but if it's true about a "better mousetrap," then Bob Noe should expect the world at his front door.

Bob and his wife, Stephanie, have a country-style life on their horse ranch in southern California. They love their horses, fresh air, and the sun. But with the good life, there's always a "fly in the ointment." And this couple's fly wasn't a fly; it was a rodent. Gophers, rats, and mice were ruling the roost. Something needed to be done, because these two lawyers hate rats. (Insert lawyer joke here.)

Bob tried everything on the market to rid their ranch of the pests. Nothing worked. And it seems the Noes' neighbors tried as well. Bob and Stephanie lost their beloved Labrador to rat poison

that their neighbors had put out. Soon after that, Bob watched an owl die and, suspecting poison, had it tested. He was right. He was witnessing this rodent fight getting out of hand. The whole pest thing kept gnawing at him.

As a young boy, Bob watched his dad and uncle invent and manufacture the first seat-belt. He then saw them sell their auto-changing safety invention to clients such as the Ford Motor Company. Bob figures he must have inherited that inventing gene.

One day while on a routine fence check, Bob saw one of those pesky gophers meet his end—electrocuted at the end of an exposed hot wire on his horse fence. That sparked an idea for this "gentleman cowboy," as his friends

Rats check in but they don't check out.

call him. After months of trial and error, he invented the Gopher Zapper.® He started AgriZap, Inc. to produce it.

Bob explains how the Gopher Zapper became the Rat Zapper. "A friend who administered property for John Hancock had a rat in the building that he couldn't get rid of. He asked me to turn my gopher trap into a rattrap. I'm a tinkerer, so I converted the Gopher Zapper to see if it could exterminate rats." It sure did. So Bob turned the Gopher Zapper into the Rat Zapper.

"I sat back and realized the potential of this thing. There are tons of mice and rats out there. And they have a heavy impact on society and people's lives. When I looked at the competitive landscape, I saw poison, gory snap traps, and inhumane glue strips that keep the animals in place until they starve."

The mouse and rat population seemed to be a better group to go after than gophers. "A lot of folks don't have horse sense, so I didn't pursue the gopher trap and its dirty, damp environment. I determined it would be better to go for an aboveground environment. My objective became creating the best mouse and rat-trap that had ever been conceived. No improvements had been made on trapping mice and rats in the last century. We can build a better mousetrap. One that is efficient and humane."

People thought he'd been zapped of his good sense. He had worked as corporate counsel for a large NYSE company for thirty years. And this is how he was going to spend retirement? "A lot of my friends and associates thought I was crazy. Now they're not laughing."

Unsolicited testimonials began to pour in; you can catch them on the Rat Zapper website. "It's gratifying to see that people buy it and then buy more. They've found that it works!"

It took countless failed prototypes to get to the profitability Bob enjoys with the Rat Zapper today. He's been at this for fifteen years now, and spent millions of dollars of his money as well as his friends' and family members' money. "I'd love to have done it overnight. When I look back, I think, 'Why didn't I do it that way at first?' But progress takes time. In the end, I think all the failures made it better. We had an odd development. Without corporate backing, research and development always takes on a different complexion. We weren't a large corporation with a big lab embarking on yet another project.

"The difference is passion. If you believe you can come up with something that's dramatically better—when it becomes part of you and you just stick to it—you'll come up with a superior product. Be patient. Don't give up too quickly."

He's following his own advice with his latest challenge: a patent lawsuit, initiated from a company that had posed as a partner. "It's sad. After you go through all those challenges and your business is going along well, then a large corporation goes after you."

But neither Bob nor his wife, who has worked as a correspondent for PBS's *Life & Times,* are mousy about it. They're bringing their attorney training and connections to the case. "A lot of people get ground up and spit out early in the legal process. We should have been extinguished months ago. They're trying to spend us into the ground. But we're unique in that we have law backgrounds and are feisty.

"Patents are not self-enforcing—that means patent holders have to enforce them and be prepared to pay gobs of money, even bring the dispute to court. In the patent arena, it takes $1 million

just to get to the courtroom. Our case is really nasty because they were partners. In the 1950s, my dad did business with a handshake. There's a real loss of integrity today."

It looks like they're going to have to rat on their former partners. "Fortunately, we got records from their manufacturer. From day one, they were using our breakthrough product as research to test market, intending to replace it with their own."

While they're hemorrhaging money in legal expenses, AgriZap, Inc. keeps expanding, in spite of the stolen sales. The company sells thousands of units each month. It was one of the top-selling products in the *SkyMall* catalog last year. AgriZap's Internet business has grown from 3 percent of total sales a few years ago to almost 50 percent today.

But that's not the end of this tail. In addition to selling the Classic and Ultra Rat Zapper, one of Bob's products is called the Rat Tale™. A plastic mouse, it connects to a Rat Zapper. Its blinking, beady eyes tell you when you've got a living mouse or one that's dead. One of his customers who has a manufacturing warehouse put the Rat Tale high enough for his employees to see. He pays a five-dollar bounty for the first person to call maintenance to respond to the blinking lights! Much cheaper than hiring an exterminating company.

It seems Ralph Waldo Emerson was right. If you build a better mouse, or rat, trap the world *will* line up at your door. Bob and Stephanie Noe took that spark of an idea and turned it into a shockingly profitable product.

Enotrab™ Optical Alert Stand

Can You Hear Me Now?

Jeff Bartone's creative inventor juices started flowing as a child. He was always fixing things. "We weren't poor, but very budgeted. In my family, if you wanted something, you had to make it yourself, find it, or fix it." Jeff grew up always looking for the next new thing, seeing parts as possibilities, and watching for new ways to combine them.

As opposed to most of the other inventors I've talked with, it didn't take a problem for Jeff to come up with an invention. The idea for the Enotrab Optical Alert Stand seemed to come out of thin air. "As an inventor, I look in the oddest places for my next inspiration. One of the best places I've found is a junkyard. My wife, Paulette, and I saw a dog-ripped sneaker blinking when a pile of garbage fell over. We grabbed the sneaker and took it apart to get at the flickering light. There had to be a problem waiting for this solution."

Jeff and Paulette had cell phones on their mind, since they'd recently invented the Pop A Lot™, a pop-up cell phone magnifier that automatically pops up when you open your flip phone. It boosts your display screen by 2× magnification. They came up with it in early 2000, when cell phones were coming out everywhere. Jeff joked with his friends about cell phone wrinkles resulting from people squinting to read the tiny screens. The problem hit closer to home when his parents complained about having a hard time reading the small print on their cell phones.

Again, his inspiration came from the oddest place—he was admiring his daughter Alyssa's

STAT BAR

PATENT: information not provided

PRODUCT PRICE: $12.99

STATE: Connecticut

INVENTOR'S AGE: both in mid-30s

INVENTOR'S PROFESSION: inventor

MONEY SPENT: "too much"

MONEY MADE: "not enough, so far"

WEB ADDRESS: b1p.info

"We do a lot of research to see if an idea is worth bringing to the market before we patent a product."

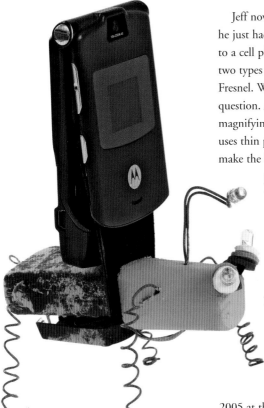

Jeff now had a magnifying solution—he just had to figure out how to apply it to a cell phone. He learned that there are two types of magnifiers—optical and Fresnel. What's the difference? Good question. An optical magnifier is like a magnifying glass. A Fresnel magnifier uses thin plastic with grooves in it to make the image larger. For the cell phone issue, Fresnel was the way to go, because optical magnifiers are too thick. The next step was finding a way to get it to pop up.

Once again, Jeff's inventive mind sprang into action. His solution was—springs. "The hardest solutions to come up with are the simplest," Jeff explains. It won the Connecticut State Innovator's Award in October 2005 at the Yankee Invention Expo and is Jeff and Paulette's biggest seller.

Back to that flickering light at the junkyard sparking an idea. The blinking light from the sneakers only flashed when the sneakers moved. They had the wiring; the question was, how to apply it? Then, like a flash of light, it came to them. A visible cell phone ringer! The solution now had a problem: cell phone users who are in an environment where they need to have the ringer off, but want to be alerted to an incoming call.

Problem solved. Next question: "What do we put our bright idea into?"

They went through a dozen prototypes to tweak the housing problem. They wanted their cell phone companion to be cute and adorable. Different versions included the Crushed Man that looked like a squashed man when the cell phone was placed on it. Cute, but they had a hard time getting his pants to look professional. They dubbed another prototype the Mohawk. He got his name from the fiber optics they used in place of an LED antenna. He was also cute, but not practical. The fiber optics were expensive and difficult to cut to the size needed. They'd lose hairs all over the place—some of the guys had only three hairs left. Another prototype was a robot whose eyes lit up. It would dance around on your desk when the phone rang—but it wouldn't stop. You had to catch it to turn it off.

Through trial and error—and endless creativity—they finally came up with a lovable bug they called Enotrab. (Where did they get the name for their cell phone bug? I'll tell you that a little bit later on.) As for the bug, he's a hit. People respond to it. The new model has a motor that powers a tongue that looks like it's catching a smaller bug. Its teeth even chatter. The Bartones manufacture the quirky Enotrab at their home in Connecticut.

artwork on the refrigerator. It was held in place by a plastic-sleeved magnet, framed with: "Everybody can be a daddy, but it takes someone special to be a father." It was an endearing message . . . and apparently an inspiring one as well. While gazing at it, Jeff noticed the corner magnified her artwork. Ring, ring, there's a clever idea on line one.

They sell the Enotrab Optical Alert Stand, as well as their Pop A Lot and Macro Lens for camera phones, on their website. Big companies have expressed interest in their products but, so far, sales have been sluggish, because they haven't started to actively market their products.

Jeff and Paulette, two petite and young-looking inventors in their mid-30s ("People think we're kids. We're not kids, we have kids!"), have officially been in the inventor business for one year. And in that short time, they've brought three inventions to the market. In the twelve years they've been together, they've always worked for themselves— first with a window cleaning business, then as professional painters for seven years. They were the busiest painters in their town, but while painting, they kept their inventor wheels turning. When they'd saved some money, they patented one of their ideas—the Pop A Lot Cell Phone Magnifier.

That was the beginning of their Bart 1 Products business. The name "Bart 1" is a take on their name "Bart-one." Get it? Clever couple, huh? Jeff and Paulette even sold their house to finance this venture. Jeff says, "So far, we've doubled our stress and halved our money. But that's a chance we had to take. We were tired of saying, 'I could have done that.' A husband and wife business is very difficult. We're running the business day and night. It's tough to market our own inventions, and manage two kids and two dogs at the same time."

Besides investing their hearts and souls, they've invested their hard-earned cash. "It's hard to say how much money we've invested. It seems like a lot. It requires continuous time and money— we're always pushing the product. When we make a couple of dollars, we put them back into the business." They say on their website: "You can't eat proto-types, and unfortunately the bank doesn't accept homemade money. It's probably a bigger risk than the lottery. It's like buying tens of thousands of tickets—hoping you'll be the one."

They still have a lot more ideas; after all, you can't stop those creative juices. But for now they are selling just three products on their website. Until they get a return on their investment, no more products for the Bartones. "We do a lot of research to see if an idea is worth bringing to the market before we patent a product."

This creative couple is looking to find a niche, maybe working with an existing company in its research and development department.

Jeff and Paulette will keep pushing the Enotrab. They don't mind being a pest if it means every cell phone has their bug.

Oh yes, and about the name Enotrab? Well, it's their last name, Bartone, spelled backwards. I told you they were clever.

Ride 'em cell phone

Backlit Paper Cutter™ / Hover Pad & Bench™ / Sea Stones™

You Light Up My Paper

STAT BAR

PATENT: pending

PRODUCT PRICE: "check at Staples for paper cutter"

STATE: New Hampshire

INVENTOR'S AGE: 42

INVENTOR'S PROFESSION: mechanical engineer

MONEY SPENT: more than he's made—(no comment)

MONEY MADE: (no comment)

WEB ADDRESS: hoverpad.com and sea-stones.com

"So I sent in my invention and expected nothing . . . "

In addition to bills, you never know what else the postal carrier might bring. Most of the time we get a bunch of ads, which get tossed into the trash can. But every once in a while, a mass mailing hits just the right person and it changes that person's life forever. That's what happened to Arra David.

When Arra opened a flier from Staples, he found the office supply giant was looking for inventors to submit ideas for original office products. The 42-year-old mechanical engineer realized he might have a winning idea. A couple of years earlier, he had created something that can only be described as a backlit paper cutter.

"When I had invented the backlit paper cutter, I had been thinking only of solving my own problem.

"It was on my third trip back from the printer to the paper cutter one day when I finally got frustrated," he explains. "When it comes to using a paper cutter, you have two choices: creep up and take a nibble at a time, or take the 'go for it' approach and give it your best shot." To stop having to reprint and recut things, Arra installed a light onto his paper cutter so that he could see precisely where the paper would be cut.

Although Arra had been using his invention for a couple of years, he hadn't ever thought of bringing it to market until the Staples Invention Quest contest came along.

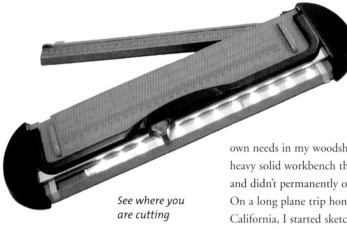

See where you are cutting

"So I sent in my invention and expected nothing. And then I heard back, 'Congratulations! You are in the finals!'"

The backlit paper cutter had been selected from 8,400 entries. Arra and his wife, Suzanne, were transported from their New Hampshire home to New York City, complete with an appearance on national TV.

Since then, the backlit paper cutter has changed a lot. "It's now totally portable," and has new features including the ability to cut decorative scalloped edges, an idea which came to Arra from scrapbookers.

With his backlit paper cutter now licensed, Arra has turned his attention to his other inventions, which have also been born out of solutions he's found for his own personal use.

Hover Benches and Pads, for example, says Arra, "came from my own needs in my woodshop. I wanted a heavy solid workbench that didn't jiggle and didn't permanently occupy a spot. On a long plane trip home from California, I started sketching," borrowing a concept from air hockey tables.

When Arra got back to New Hampshire, he faxed his design to his dad, Raffi, also a mechanical engineer and "the smartest man in the world." Arra asked his dad, "Explain to me why this won't work," because if it would work, surely someone would have already done it.

The design did work, Raffi concluded, and so father and son built the

This heavy, solid workbench floats on air.

bench together. The result is the Hover Bench, a "heavy solid workbench that's totally portable."

Sticking true to his advice to always be open for suggestions, Arra soon realized that people had other heavy items besides workbenches they'd like to move. This realization gave rise to the Hover Pad, a 1-inch-thick pad that you can place under any heavy object, and easily move it around. Arra says he has since sold "hundreds" of pads and benches.

Another idea came along when Arra and Suzanne were building their dream home and needed a door handle for their mudroom. While at a beach they picked up smooth stones, and then installed them around their home as door handles. Family and friends were soon asking for sea stones and Arra's entrepreneurial mind kicked in. Now Sea Stones are sold nationally and internationally.

Arra says he is concerned about the environment. That's why his self-designed and built "dream home" is heated and cooled without any oil or natural gas, via a system of geothermal heat pumps. And whenever he takes sea-smooth stones from a beach to use for their business, he always "re-plants" the beaches with rough stones so that those stones will eventually smooth out and replace what's been taken.

World System Clock™
Time for a Change

"The biggest challenge has been making the clock reasonably priced . . . "

You could say that the idea for the World System Clock had always been ticking around in Werner Schulz's mind. "I can't recall when exactly I first had the idea for the clock," which shows the time in all twenty-four time zones around the world at once.

"I grew up in Switzerland, which is watch country. There were three different watch companies on my street alone," says the 77-year-old who moved to the United States in 1951 to study and pursue a career in graphic arts. "I know I carried around the thought about the clock with me for a long time."

Yet it was time, or really a lack of it, that for many years kept Werner from creating his World System Clock. Instead, the would-be inventor pursued a career as a graphic artist for Dixie Cup Company.

But while he didn't yet have the time on his hands to develop the World System Clock, Werner used his time on the clock at work to help invent another timeless item—Dixie's famous Riddle Cups, featuring a different riddle on each cup. Until he came along, the cups had always featured the same pattern on every cup. What a bore. No need to make this a riddle. Werner's cups were a huge hit. "The new Riddle Cups sold like cupcakes—about $40 million worth, which was about four times as much as the previous design."

When Werner retired from Dixie in 1988, the time was right to work on his clock, but it was really in the last three to five years that the invention finally started ticking along.

At his Newtown, Connecticut, home, the married father of three and grandfather of two designed a system that displays the time in all twenty-four time zones simultaneously via a double clock and double wristwatch. The basic idea of the World System is to replace the hour hand with a disc, showing the time in twelve cities, with the owner's home city specially denoted. At only a second's glance, the user of the clock or watch can read the time everywhere from Anchorage to Zurich.

While waiting for his patent to be finalized and looking for a company to manufacture and distribute the timepieces ("Timex would be perfect"), Werner has time to envision a big future for the World System Clock. "I've talked to pilots and there's nothing like it on the market. They love it—and they especially love the fact that I've put a red triangle on the clock's hand indicating Greenwich Mean Time, which is what pilots use."

Werner has invested only "pocket change, a hundred or two hundred dollars, plus time and effort," in the invention process so far. "The biggest challenge has been making the clock reasonably priced so that everyone interested in it for educational purposes can own one." He feels strongly about its educational value. He says, "Children can easily learn that we have twenty-four

Time traveler

hours in a day and twenty-four time zones in this world. It will also help them understand world geography."

Werner's business card reads "thaumaturgist," which means "performer of miracles or magic feats." I'm not sure if that's an overstatement. After all, it might be debatable as to whether Riddle Cups and his World System Clocks are

truly miracles. But there's no doubt he's a bona fide inventor. After all, disposable cups will never be the same. And Werner is getting close to changing the face of clocks around the world.

So here's a riddle. When will Werner's World System Clock show up in stores? The answer: only time will tell.

The InfoCube™

Waiting for a Waiter

STAT BAR

PATENT: pending

PRODUCT PRICE: prefer not to print

STATE: North Carolina

INVENTOR'S AGE: 38

INVENTOR'S PROFESSION: engineer

MONEY SPENT: an easy $300,000

MONEY MADE: $30,000–$40,000

WEB ADDRESS: theinfocube.com

"...the whole reason I designed this was because I was impatient."

It's turned into an American dining ritual. You get to the restaurant, you give your name and the number in your party, and you're given a pager. You hang on to the pager until it starts to flash. A flashing pager means your table is ready.

Is it a perfect system? Definitely not. Can it be improved? "Yes" says David Thompson.

He describes his invention epiphany this way: "I was waiting for forty-five minutes to be seated in a restaurant and became bored after the initial talk with my companion had expired. Holding the coaster-style pager, I thought, 'Boy, I can do better than that. This pager could be entertaining me while I wait.'"

He looked into the available technology and brainstormed different entertainment solutions. He knew it had to be cost-sensitive. He figured the restaurant industry wouldn't spend a lot of money on pagers. So his pager had to be rugged as well as cost-effective. As an electronics engineer, Dave was able to design the entire unit himself.

"I didn't make prototype after prototype. I used one and developed it a piece at a time until it worked. It's like making a car—you don't throw it away because the back end doesn't work. You work on the back end until it's right." He's an engineer, after all. Once he knew he got his invention right, he dubbed it The InfoCube.

Dave tested out his InfoCube at a local restaurant. He says the restaurant owners and patrons loved it. Dave's InfoCube let hungry soon-to-be diners look

While you're waiting,
here's your Infocube.

over house menu specials, play games, get news, trivia, sports scores, movie times, fun facts—and even order drinks and appetizers while waiting to be seated.

Both the restaurant owners and customers liked The InfoCube, because it made the wait for a table seem shorter. Amused waiters (by that, I mean those waiting for a table, not those serving meals) equal happy diners, and that takes stress off the host or hostess.

Other perks of the The InfoCube include statistical information. Management could use the cube to track average waiting times, number of pagers used, and number of walkaways. If patrons placed a drink or appetizer order while

waiting, it kept them happy and reduced the time they ultimately spent at the table. Now that's good use of information.

After the restaurant success, a local Toyota dealer sought out Dave's invention. They were using 1,000 pagers a week to notify customers their oil change was finished or their car might need more maintenance than they'd bargained for. The pager signaled them to come in and have a chat. Ultimately the dealers sold more air filters, the service department ran more smoothly, and customers weren't bored while waiting. That's what I call a win-win, cubed.

Dave has learned what many inventors eventually learn. "Unfortunately, it takes more than a great idea to sell an invention. It takes a good marketing strategy and competitive production. I produce The InfoCube in the U.S., but I'm looking into producing it overseas to get a better profit margin. That way, I can attract more distributors."

He's currently promoting The InfoCube himself. "I have a couple of distributors and dealers, and I'm trying to contact people in the industry. I wish I

had an advertising budget and could travel to promote it. I go to a show once a year, the National Restaurant Association" (NRA—think food, not guns). Dave is now looking for capital investors to get his pager shaking. Check, please.

The InfoCube is all about making a wait tolerable. Now, Dave is stuck playing the waiting game. "If the response on The InfoCube was so-so, I'd be questioning its potential. But with such a positive reaction, I'll keep going. I'm impatient; I want it to explode. I want to sell thousands right away. Come to think of it, the whole reason I designed this is because I was impatient."

Dave has been pushing his cube for only about a year, but he's found that it's difficult being the new kid on the block. "A lot of people don't have confidence buying technology from a new company. They don't know if you'll be there in a year for tech support. It has cost me $300,000 to build up my inventory, so it will take quite some time to break even, let alone think about drawing a paycheck."

Dave is still busy cooking up other InfoCube applications. "Pagers are used in churches, daycare centers, car dealerships, restaurants—anywhere people are waiting, owners can use a pager to let customers know that what they're waiting for is ready."

Dave, your table is ready.

Core Sound Trees™

Tree-mendous

"People like that it's fireproof and easy to set up."

You probably don't know that imported fish are drowning the U.S. commercial fishing industry. With lower labor and fuel costs, farms in India can raise fish cheaper. They're cutting prices, so U.S. fisheries are cutting jobs.

Neal Harvey is all too familiar with the situation. Born and raised in a commercial fishing family, he's been operating Harvey and Sons Net and Twine now for thirty-five years. His family-owned business makes crab traps or "crab pots" as they're called locally, as well as nets, and trolls. He didn't sell enough nets or trolls last year, so he reeled in that part of the business. Even his crab pots aren't moving. It's Core Sound Trees that keeps them afloat now.

Core Sound Trees were created when Nicky cast his net for a new kind of fish. At 62 years old, he was looking for a product to keep the business swimming for his sons. He got a creative bite when, high up on a shelf in the store, lights strung on a bit of wire caught Nicky's eye. That's when crab traps became Christmas trees and Nicky, sort of, became St. Nick.

Back in his workshop, not Santa's, Nicky played around with the wire mesh scraps left over from making crab traps. One thing led to another and the first Core Sound Tree was erected. It was only 18 inches tall. People liked them, so Nicky made more in different sizes.

The trees were bulky at first. That was okay until he made bigger trees—they were hard to transport in people's cars. Nicky cast about for another idea and designed a way to flatten the three-foot-tall trees. Once he figured that out, he quickly graduated to six-foot trees. It's the Core Sound Tree's collapsibility that Nicky ended up getting patented.

The built-in lights became an afterthought. You see, the women who were decorating the trees had to remove all of the lights before folding them up. It was a pain, so they asked Nicky, "Why don't you attach lights so it's all ready to go?"

Nicky didn't like the complaints, but he liked the idea.

The trees are handmade by crab pot makers out of coated crab pot wire that's made to go in saltwater, so they'll last a long time. What started out as a yard ornament is now being used inside. "People like that it's fire-proof and easy to set up. The lights are already on. They just have to put their decorations on, then fold them up when they're done." They've been making the trees with green wire. But now, in response to a request from customers up north with snow, they'll

Christ-mesh tree

offer Core Sound Trees in white with white lights.

The entire community has been behind Nicky in his new adventure. He says, "I've been making commercial troll supplies all my life. People told me, 'You better get a patent on those things before people start taking your idea.' I'm glad we did, because they really took off."

The first year, a few neighbors wanted them. The next year, he sold about ninety and has doubled sales every year since. "We didn't step out too far until we got a patent on it. That was a year-and-a-half process. Our patent pending was the go-ahead to venture out of the county and sell to other dealers." The Core Sound Tree is now sold through

more than forty dealers up and down the East Coast, from New Jersey to Georgia and inland to Kentucky. He has Internet buyers covering the rest of the coast, from Long Island to Florida.

To keep up with demand, Nicky and his crew have had to start production a month earlier each year. They used to start making the Core Sound Trees when crab pot season trailed off in September. Now, they're starting the trees in mid-summer to have enough in stock.

The Core Sound Trees have been a sound investment for this fisherman. "We made more on the trees in three months than on crab traps all last year." Sounds like he's hooked quite a catch!

Who said only God can make a tree?

STAT BAR

PATENT: US #5377739

PRODUCT PRICE: $19.95

STATE: Dallas, Texas

INVENTOR'S AGE: 47

INVENTOR'S PROFESSION: senior contract administrator for aircraft sales and marketing

MONEY SPENT: $10,000–$15,000

MONEY MADE: not a thing

WEB ADDRESS: inventionconnection.com/BOOTHS/booth172.html

"...she made sure it was me by putting her index finger through the mini-blinds."

Peeping Thomas™
Tom, Is That You?

Having a stranger break into your home is a universal fear. We buy locks, timers, and alarms to prevent a break-in. It's not only having your possessions stolen but also having uninvited strangers snooping through your home that violates your personal space—and frightens you.

Terry Kirby, a 47-year-old who works in contracts and legal services for the aviation industry, knows this fear firsthand. He'd bought his dream house, ignoring his friends' and family's warnings and the neighborhood's crime statistics. He thought that nothing would happen to him—until he walked in on three burglars robbing his home.

On edge and unable to relax, Terry drove to his grandmother's to sleep there. "Before she opened the door, she made sure it was me by putting her index finger through the mini-blinds." Then Terry's grandmother, Mattie Lee Davis, joked that she hoped she didn't look like a Peeping Tom.

"That night, I remember thinking what a brilliant little invention having your own Peeping Tom would be.

Neighborhood Crime Watch Programs would definitely support it, and so would local police departments."

Terry wondered how to make his Peeping Tom idea work. It had to be small enough to fit most windows, and have a small gear, a tiny motor and a relay switch to allow it to operate in 5-, 10-, 20-, and 30-minute intervals. A bonus would be a setting that would randomly lift and lower a single mini-blind slat.

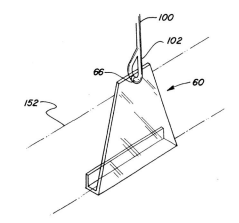

Terry had the background to make this invention work. Terry went to work for Texas Instruments right out of high school. From soldering circuit boards together, he moved up through the ranks as a mechanical engineer. If he could figure out the inner workings behind the missiles the government paid Texas Instruments to build, he could figure out the electronics for his home security idea.

The spark he needed was kindled one night while out to dinner with friends. Having kept his losses and safety fears inside, Terry was embarrassed. He excused himself, only to let tears spill out in the restroom. Then, as he washed up, he heard a hissing sound, like a mother soothing a child. He looked up—at an automatic air freshener. "Whatever made that device spray was what I was looking for to use for the inner workings of the Peeping Thomas. I bought one, took it home, and tore it apart. I created my prototype and had my Peeping Thomas invention patented."

Terry spent $8,000 on the patent, and lost a bundle on a marketing rip-off from a company that promised him everything. He was certain that after getting his patent, this marketing firm would seek a license to have it manufactured. After spending all of his savings, Terry hasn't been eager to invest any more in the Peeping Thomas, though he still believes in his product.

Others do, too. The Lifetime cable network called him about a new television program with a segment featuring inventors and their inventions. "I was scheduled for the next broadcast, but they didn't have the ratings to keep the show going." A company in Florida wanted to make three commercials for him and air them in the Ohio Valley. The company was looking at a licensing agreement, but Terry didn't want to spend the $5,000. QVC was interested in it, but it would take $30,000 to $50,000 to do the tooling to have a product to sell. Terry doesn't have that up-front money. His contact at Wal-Mart said it's not about having it manufactured; the key is developing a marketing strategy that creates consumer demand.

Terry wants the world to know about his Peeping Thomas. So far, with limited funds and no advertising, there hasn't been a peep about it. Terry hopes that will change. Terry's twin brother, Jerry, dedicated to creating that much-needed buzz, has designed a website. The idea is to explain the Peeping Thomas concept to potential licensers and maybe show them a demand for the product based on website hits. Then Palmer, Terry's wife, would take that data and create a product release. She's a prototype brander and advertising guru by trade, perfect for this part of the game plan.

While the product is not yet for sale, the patent is—for $500,000. But Terry says the buyer doesn't have to pay that up front, just sign a licensing agreement. The royalties come off the top, so the buyer gets his or her initial investment back, and he gets his piece. Terry figures the Peeping Thomas will easily sell for $19.95 and he's priced production of a small run at less than $3 a unit. So Peeping Thomas, aka My Nosy Neighbor, should be able to turn a profit. Any investors out there interested in helping Peeping Thomas, or Nosy Neighbor, scare away the bad guys?

Someone is watching you.

Compact Disc Eraser™

Groovy Solution?

STAT BAR

PATENT: pending (10-951,952)

PRODUCT PRICE: $14.99

STATE: California

INVENTOR'S AGE: 32

INVENTOR'S PROFESSION: entrepreneur, former engineer ("engineering is not as exciting anymore")

MONEY SPENT: $2,500

MONEY MADE: $500 in first month on the market

WEB ADDRESS: disceraser.com

"This is the only data security tool that promotes recycling."

Wade Sun has burned hundreds of CDs and DVDs, backing up a lot of data in his seven years as an engineer in the storage technology industry. But what do you do when you need to erase that information?

Wade looked for a simple way to destroy discs and the confidential material on them. He tried breaking them in half, hammering them, sanding them, and cutting them up. Sure, he was able to crack the CDs, but was left with sharp shards that this father-to-be didn't want around the house. Wade even invested in expensive CD shredders, which would work fine for the large volume he has, but what about people who only have a few CDs they want to destroy? "My original thought was, 'Is there any way to disable a disk without physically destroying it?'"

Wade's research on ways to destroy CDs took him to the flip side—ways to restore them if they get scratched, such as polishing kits. He asked himself, "Is there a way to scratch a CD so it goes beyond polishing?" Wade did more than scratch the surface. He came up with a new method to put grooves into the disc. He calls it "optical strip technology" and uses it in the Compact Disc Eraser he has invented. The Disc Eraser is a small, easy-to-use, economical device that puts a series of deep grooves in any unwanted disc.

To see if his invention worked, Wade took his Disc Eraser to a data recovery company that specializes in recovering scratched CDs. "They're the experts. I

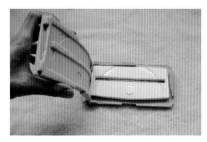

took my product in to the president. Impressed, she said the scratch renders the data practically unrecoverable—and she'd never seen anything like it. She especially liked its small size.

"I've gotten great feedback from businesspeople, too. They can carry the Disc Eraser around in their briefcases. It's hard to carry a shredder around. And my product promotes recycling. CDs and DVDs are made of recyclable plastic, but recyclers won't accept broken glass or plastic. This is the only data security tool that promotes recycling. One-half million tons of CDs get tossed every year, so people should be recycling them, just like plastic, paper, and glass." This recycling benefit is one of Wade's strongest marketing angles.

He has had start-up help. His Disc Eraser became a student project for an MBA class at San Diego University, where Wade's wife, Maria, is a graduate student. The class provided Wade with a detailed

business plan and cost estimates. Wade has gotten support from his dad, a small-time inventor in Taiwan. "He coached me to keep the design simple and easy to use. He made a lot of suggestions, as well." A marketing director for a catalog company put in his two cents. "He gave me suggestions for setting the price point and designing packaging that will sell effectively. It's hard to find help from entrepreneurs because they're busy. So I did a lot of searching on the Internet and learned from various experts. "I'm an engineer by training. Business marketing requires using the other side of my brain."

That other side of his gray matter is paying off. His first month out, he sold thirty Disc Erasers through his website, eBay, and Amazon.com at $14.99 each. It didn't hurt to be in the top 30 out of 10,000 inventors for ABC's *American Inventor* TV show either. "I got to demonstrate my product on episode five. It helps my marketing to be on TV, even if only for ten seconds."

He eventually wants to get the Disc Eraser in retail stores, such as Staples and Best Buy. "I want to outsource the manufacturing to China, but we're not quite ready to produce a high volume. We're still refining product details." For now, he's packaging in the U.S. It's expensive, but he's able to watch those critical quality-control issues. "I only package the good ones."

If Wade keeps this product spinning, he should start to see some profits. "I hope to break even soon. It's a struggle. A lot of inventors don't have the resources and hire experts to do the work for them. People have worked ten years on the same thing and spent over $100,000."

Not Wade. He kept his costs down by doing his own patent application, website design, and marketing research. And if his Compact Disc Eraser stays on track, he should soon be able to erase his debt and turn a profit.

Made by scratch

Conedoms
Cup-a-Cake
E-Z Lift Turkey Transport
Marshmallow Chef Sticks
Octodog
Grandpa Witmer's Old Fashioned Peanut
Butter Mixer Crank
Cookie Stacker
Mac & Cool / Potty Mitts
Bottle Sling
Cozycooler
Motorized Ice Cream Cone
My Pet Fat

What Goes In . . .

Eating—we are obsessed with it. We Americans love our food. If you asked me if we eat to live or live to eat, I would confidently say that most Americans live to eat. There are thousands of cookbooks and even more diet books filling the shelves in our bookstores and libraries. As a nation, we are overweight. In our society, chefs have achieved celebrity status equal to rock stars. There's even an entire cable TV network devoted solely to food. Food is the glue that holds together every family event, every social gathering, and even every business meeting.

Our preoccupation with food has given inventors something to chew on. Our love of food has spilled over into a love of gadgets associated with food. From a dish that cools off a hot plate of macaroni & cheese to a motorized ice cream cone, this batch of

inventors has found products you can really sink your teeth into. Do you need a hand carrying a cooked turkey across your kitchen? Have you ever thought your hot dog was boring and instead wanted it to look like an octopus? Maybe a yellow blob that looks like five pounds of fat could be your next diet buddy? You might never have asked yourself these food-friendly questions, but these inventors have.

So put down the menu, walk away from the refrigerator—if you are hungry for food-related gadgets, I'll serve up a few. Some might look at these inventions as exploiting our love of food; others might see them as a way to savor that love. Without going off half-baked, I can safely say these inventors all cooked up innovative ideas, which were certainly food for thought.

Conedoms®
Safe Sex for Cones

... the cone stays free from germs and the cone lover stays free from drips.

Meet 65-year-old Bob Sotile. Some might consider him the safe-sex advocate for ice cream cones. Whatever you decide to call him, he's one smart cookie or, in this case, maybe it should be one smart cone.

Here's the scoop. The first thing you should know about Bob is that he loves the chase. In fact, he says he likes the chase more than the money itself. At least that's what he says. Bob got his start chasing sales in medical equipment and pharmaceutical supplies after dropping out of the University of Rochester. Selling everything from exam tables to microscope bulbs to tongue depressors, Bob's business flourished with thirty-one employees, taking in millions in sales. After twenty years he cashed in, and then sold his business to a national company.

Picture this. In 1997, on a warm summer evening in Rochester at exactly 7:00 p.m., Bob indulges himself with a cone of chocolate almond ice cream at a local ice cream shop. As he tells it, he ordered an ice cream cone and gave the "scooper" his money. Using only her bare hands (no plastic gloves), she takes the money, brushes the hair out of her face, and then picks up the cone. Yuk, call in the food police. Bob politely interrupts her: you just touched my cone after handling dirty money. Oh no, money is sterile, she says. Bob lets it slide. But you just touched your hair, he says. Oh no, my hair is clean. I just washed it. Totally grossed out, Bob says, forget the cone. I'll take it in a cup. That was when that proverbial lightbulb went off.

Bob told his wife he was going to invent a product to protect cones. She dismissed his idea, calling it nuts—and not the kind you add to ice cream. So was Bob licked?

Certainly not. Remember, Bob loves the chase, and more than that, he wanted to make the world safe for clean, bacteria-free cones and keep hands and sleeves from getting sprinkled with, well, ice cream (and sprinkles, too).

First, he tried covering the cone in paper. But that idea quickly melted. The paper stuck to the cone and was still filled with yucky germs. Bob wasn't frozen in his tracks. He settled on a completely sanitized, disposable plastic sleeve. He added a ledge around the top of the sleeve to catch any cone drippings. This way the cone stays free from germs and the cone lover stays free from drips.

Bob spent $10,000 on a prototype and started shopping it around. In 1998, he made his first real sale to Carvel Ice Cream. Conedoms were a success.

Today, Conedoms come in all shapes and sizes. Whether it's waffle or wafer or square or pointy, Bob has scooped up over a million dollars in sales (that's over 150 million Conedoms sold over six years worldwide).

And Bob's story is far from over. His son, Michael, looked at his dad's Conedoms and found a new use for them. Michael is a restaurant manager

as well as an actor (Speed, Reservoir Dogs, Miami Vice) and a former bartender. Michael recognized a problem in the bar biz. When bartenders close up for the night, they leave the booze open with just the pourer attached to the top of the bottle. This is not very sanitary. Flies and all sorts of bugs like a little nip late at night. To keep the bugs out of the booze, Michael put Cone-doms over the pourers. Bob loved the idea and started to retool Conedoms to fit the bottles even better. So the Conedoms, in a sense, gave birth to "NiteCaps™." Conedoms, NiteCaps, it all works together and I'll let you write the jokes from here.

Most of us know the importance of condoms and safe sex, but it took a medical supply salesman from Rochester to find a way to protect our cones.

A safe cone is a happy cone.

Cup-a-Cake®

To Protect and Serve

"Sometimes I wonder if we're in over our heads."

Have you ever tried to carry a cupcake in a bagged lunch? You wrap it in foil or put it in a plastic bag, and strategically place toothpicks in the frosting in the hope that it won't get mushed. You try to handle your bagged lunch oh so gently, but when lunchtime rolls around, you've got yourself one nasty looking cupcake. Right?

Well, one family with ten grown siblings sat around at their kids' soccer and baseball games talking about all the tricks they'd tried to avoid mangled cupcakes in their kids' lunches. One day, according to Colleen, one of the ten, "It's not like we sat there and dwelled on what new product we could invent. We were talking about the cupcake smear one day, and wondered why someone hadn't come up with a solution. We decided that 'someone' was us." Cupcakes would never be the same again.

Five of the ten siblings in this Michigan family—Peggy 51, Colleen 49, Jim 45, Sally 44, Bob 42, and Colleen's sister-in-law Kim 49—cooked up a plan to have their cake and eat it too. They got right to work on a prototype. "We bought cupcakes at the grocery store—the ones that come in the plastic containers holding six cupcakes. We stuck toothpicks in the plastic container to see how far our prongs needed to go to hold a cupcake in place. We tested it by turning it upside down to see if it would fall."

Once they smoothed it out, they took their idea and measurements to an AutoCAD engineer. They wanted the product to have ridges like cupcake paper, but that wasn't feasible. They did make it tall, leaving a smush-free zone for the frosting. Three prototypes

were made before they hit upon a design that worked. Then they hired a local plastic manufacturer to make the official Cup-a-Cake and manufactured it in different colors—clear, blue, yellow, pink, and green. It was time to take this party to the people.

The first stop was at Pennsylvania's Invention & New Product Exposition. They made quite a few contacts and were invited to show the Cup-a-Cake container to the Electronic Retailing Association. Unfortunately, it didn't sell like hotcakes. After talking with more of the pros, they were reassured that they had a good idea. Their break came at the International Housewares Show in Chicago—they were filmed by HGTV twice. That gave a boost to sales! And so did the Cup-a-Cake's sweet coverage on *Good Morning America*. A famous cook was showing new products for the

Protecting cupcakes everywhere

kitchen and asked Diane Sawyer which one she liked best. "Oh, I love the Cup-a-Cake!" Diane declared. Doesn't that just take the cake? "We got a lot of action on that one," Colleen said.

The Cup-a-Cake has been featured in more than twenty-five newspapers, including the *Wall Street Journal* and the *Chicago Sun-Times*. Articles were also in *Parents* magazine and *Good Housekeeping*. These kept their website sales cooking. But the Cup-a-Cake website is not their only sales gig—they've gone retail. Cup-a-Cake is sold in bakeries across the U.S. and Canada through *The Baker's Catalogue*, and in numerous stores including Kiss the Cook, the Container Store, and Harry and David.

The five are getting enough bites to feel that this is something to keep in the oven. Still, they've all kept their day jobs. Colleen is a respiratory therapist by day and a Cup-a-Cake maker by night. "We're all busy, working our regular jobs and raising kids. We don't have spare time anymore—any extra time is spent on Cup-a-Cake." They fight a lot—after all, they are siblings. Still, when the patent holders have a major problem and their discussion oozes from business meetings held at Colleen's house into family get-togethers, the whole family gets in the mix.

Colleen says launching the Cup-a-Cake has cost a lot of time and dough—a couple hundred thousand dollars—and adds, "I might have been able to pay for college for one of my kids."

"Sometimes I wonder if we're in over our heads. It's expensive to get into stores—you have to pay for your own displays. And it takes a whole lot of knocking and re-knocking on doors to try to get them open. We'd really like to license it out or for someone to buy our patent. Seeing a profit—that's the icing we want to taste."

Colleen and her family prove Cup-a-Cakes are a sweet business, but no cakewalk.

E-Z Lift™ Turkey Transport

Turkey to Go

STAT BAR

PATENT: pending

PRODUCT PRICE: $14–$18

STATE: California

INVENTOR'S AGE: "too old to care"

INVENTOR'S PROFESSION: consultant

MONEY SPENT: $200,000

MONEY MADE: not yet break-even, licensed it a year ago

WEB ADDRESS: e-zturkeylifter.com

"We filled the bathtub with warm water and put in the turkeys."

Every time the Klein family had turkey dinners, Bruce ran into the same problem. As the man of the house, he had to get the cooked bird from the oven to the carving board. "It's a macho guy task—we have to do the carving. But I couldn't find a utensil or gadget for the job."

He tried two forks—they bent. He tried a fork and a knife—the fork bent and the knife cut him. Oven mitts were a disaster, the meat stuck to the cloth. Bruce then used the ultimate macho tool: bare hands. "I burned my hands. It was the height of stupidity." So who's the turkey now?

Bruce bought a utensil shaped like a big U with a rubber handle. Supposedly designed for the job, it forced the turkey into a vertical position, so the stuffing fell out before the turkey dropped on the floor. It was time to talk turkey. "I'm a smart guy. I have some engineering degrees. I told myself, 'There's got to be a better way.'" Bruce went to work finding it and eventually he ". . . came up with a design

that was logical and seemed to work."

Bruce had samples of the first E-Z Lift Turkey Transport made from different materials, experimenting with what could best hold the bulky weight. "Some of these turkeys are like small humans at thirty pounds. My wife, Karen, told me 'There's no way a normal woman can lift a twenty-pound turkey with one hand. Why don't you design it for two?'" Bruce redesigned the handle for two hands and had Karen put it to the test. She was able to pick up a twenty-eight-pound bird using his E-Z Lift Turkey Transport.

A visiting 11-year-old girl was able to pick it up, too, and walk all over the place with it. "I'd solved the problem of moving a turkey. Now I wanted to know how far

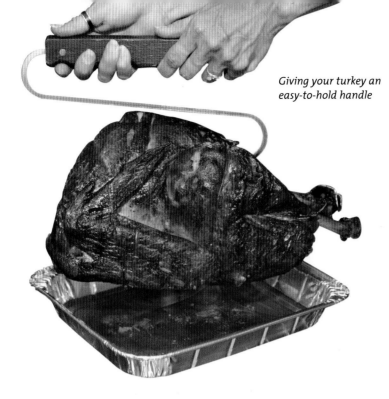

Giving your turkey an easy-to-hold handle

you can carry a turkey using this thing and not be uncomfortable. We actually had people walking down the street with the turkey." Why didn't they take pictures?

Here's how it works: The metal point of the E-Z Lift goes into the opening of the back of the bird. The point digs into the bird's ribcage, so you're lifting the skeleton. Bruce uses square metal stock because it has points on the corners. With the points, the E-Z Lift wedges into the bird and keeps it in place. Using round metal would cause the bird to rotate. Once a bird that size starts rotating, you can't stop it; it takes off and hits the floor.

Bruce had a bunch of units made off-shore. He gave them to several people he worked with and asked for their honest opinion. One guy called the day after Christmas, amazed that he didn't drop the turkey. "He said it was so easy and simple."

It was time to go to a product show. Bruce wanted a cooked turkey at the booth so people could test the E-Z Lift. He asked a woman at the hotel about having a twenty-pound bird delivered to the booth each day to use for a demo. After learning how much, they decided to pick up three turkeys from a supermarket. "So we won't cook the birds; what's the difference? We went to Albertson's at

8:00 a.m. The show started at 10:00 a.m. But the turkeys were frozen. They were like concrete bowling balls.

"I wondered, 'How can I soften up these things?' We filled the bathtub with warm water and put in the turkeys. The things were bobbing up and down. Throughout the day, we kept going up to the room to make the water hotter. No matter what we did, we couldn't get the things thawed. The next day, I ended up taking a shower in the middle of three turkeys."

"The second day, the turkeys were soft enough to demo on. But we didn't realize all the guts were in there." This would be a good time to point out that these guys do the carrying and carving, not the prepping and cooking of the bird. After serving as E-Z Lift demo for a time, the turkey would start to ooze, winding up its weight time in the garbage.

"Our booth was right at the entrance, so everyone who first came in to the show stopped at our booth. Each visitor had a disaster story to tell about his or her turkey on Thanksgiving Day. And there were hundreds. All I heard was, 'Where the hell have you been all my life?'"

Bruce and Karen licensed the E-Z Lift Turkey Transport to The American Tailgater Company a year ago. So far, they haven't seen gravy, but Bruce and Karen are certain their E-Z Lift Turkey Transport will fly.

Marshmallow Chef Sticks™

Sticking with Sticks

STAT BAR

PATENT: pending

PRODUCT PRICE: $19.95 for two or more sticks, $6 for personalization, plus shipping and handling

STATE: Pennsylvania

INVENTOR'S AGE: 44

INVENTOR'S PROFESSION: technical project manager for major financial institution

MONEY SPENT: "a few thousand"

MONEY MADE: "a little in the black"

WEB ADDRESS: marshmallowchefsticks.com

After a fun-filled winter's day sledding near their Chalfont, Pennsylvania, home forty minutes outside Philadelphia, Don Saul's three young daughters were gathered around roasting marshmallows. But the three girls were not using found branches and sticks to roast their marshmallows. Instead, Don had crafted special marshmallow sticks for his girls.

"What a great idea," a family friend exclaimed. "You should sell those on the Internet."

"I'd never thought about doing that," Don says. "But I always enjoyed woodworking. I'd built a playhouse for my daughters, for instance. And I realized that the marshmallow sticks would make a great gift, a perfect gift for that person who has everything."

The fire was lit. Don would stick it to sticks and would manufacture a perfect marshmallow holder. Don, a 44-year-old technical project manager for a major financial institution, would go into the stick business.

"The first question was what kind of wood would work best," Don explains. "I wanted it to be high quality, but since I hand-make each one, I wondered how I could do that work and still be able to offer it at a price that people could afford. I'm still always looking for ways to reduce the cost."

They come in 3 sizes.

After some research, Don found that furniture-quality red oak wood would work best. Now, if you're wondering if using wood is safe, keep in mind that for generations, marshmallows were roasted on wood sticks and branches. Wood is safe in this situation because the marshmallows cook long before the wood burns.

Next, Don had the idea to offer personalization as an option for the Chef Sticks. The ability to put their own names on a Chef Stick is "what really makes it for some customers."

After setting up a simple, one-page website, Don was in business. Using a bit of Internet savvy, he was able to study the HTML source documents of similar websites to see how they were structured, and then set up his own site in the same fashion. Thus, when an Internet user searches for keywords such as "roasting marshmallows," Don's site,

marshmallowchefsticks.com, comes up high on the list.

Don's new enterprise was soon scouted by a Wall Street Journal reporter who ordered some Sticks and had them tested by a troop of Girls Scouts in New Jersey for a feature story in the paper.

"The Girl Scouts loved the fact that they could eat right off the stick," says Don. Their one criticism was that at 47 inches, the stick was a bit long for their liking. "I thought about it, and I realized that it would be a good idea to offer alternative lengths, at 30 inches and 16 inches."

Looking back now, with "many hundreds sold," Don's advice to other would-be inventors is to "go with your gut. Just go for it. Others around you may laugh, but don't give in, don't be discouraged by nay-sayers. If you care about what you are doing, just keep going."

In Don's case, he was fortunate to have the loving support of his wife, even when it meant staying up late to ship out orders on a tight deadline. For Don, the Marshmallow Chef Sticks are a sweet labor of love. "I care about the customer. I care about the user's experience. I don't want to turn someone down, even if it means I have to stay up in the night" to make it happen.

"Does everyone 'need' a Marshmallow Chef Stick?" Don wonders aloud. "No, they don't need it. But a lot of people love to cook marshmallows," he answers, because he knows this product enhances their experience. "To me, each sale is a seed to grow for the future."

Will roasting marshmallows on found sticks and branches, become an antiquated activity? I doubt it. But Marshmallow Chef Sticks are a classy way to roast marshmallows, and Don Saul plans to stick to it.

Mmm . . . toasty

Octodog™
A Sea Monster that's Fun to Eat?

STAT BAR

PATENT: utility patent pending; design patents: US D461618-S and US D551407-S—Octopus-shaped food products, US D506363-S—Cutting apparatus

PRODUCT PRICE: $16.95

STATE: Michigan

INVENTOR'S AGE: 41

INVENTOR'S PROFESSION: manufacturer in auto industry

MONEY SPENT: $100,000

MONEY MADE: "Unfortunately, we are not in the black yet. We have paid back all of the family members that helped with the start-up costs."

WEB ADDRESS: octodog.net

A bizarre ritual involving an octopus on ice gave him inspiration.

I think Freud once said something that sounded somewhat like: Sometimes a hot dog is just a hot dog. That may be true, but sometimes a hot dog can be an octopus. Confused? Don't be. Just ask inventor Ed Suer, the inventor of, quite frankly, one of the goofiest gadgets I've ever uncovered, the Octodog. Ed turned a wiener into seafood.

At 41 years old, Ed lives, along with his wife and daughter, in the Detroit area. With a background in manufacturing metal models for the automotive industry, Ed always yearned to invent his own product. His first few ideas—a crayon that works like an ink pen and a trailer for kids that attaches to a bike—never made it out of his head. Then, when he saw those products on the market years later, he knew he had to go for it with one of his ideas.

What people like, he concluded, was food and kids. For food, he figured a gadget involving hot dogs would sell. Kids love them and billions are sold each year. Now he just needed a product that would cut the mustard.

On a family camping trip, it all clicked. A friend put slices into both ends of a hot dog and roasted it. The heat caused the slices to open up. Like magic, poof, the wiener became a spider. But Ed wanted something that would interest the

And he tastes good, too!

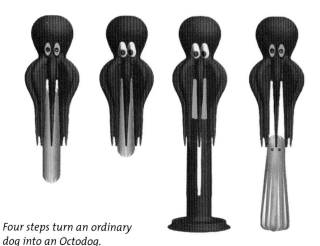

Four steps turn an ordinary dog into an Octodog.

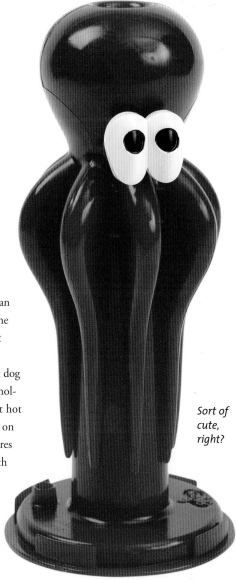

Sort of cute, right?

folks in Detroit. A bizarre ritual involving an octopus on ice gave him inspiration. You see, for fifty years, Detroit Red Wings hockey fans have been throwing octopi onto the ice after a big win. The tradition started when two fans who owned a fish shop threw an octopus onto the ice during a Stanley Cup run in 1952. Here's the part that's hard to swallow. They chose an octopus because, to them, each tentacle symbolized a win in the playoffs. You see in those days it took eight games to claim the Cup. I digress; let's get back to Ed.

Relish this question: how do you turn a hot dog into an octopus? Ed started slicing and dicing until he found the right formula. Using his manufacturing skills, he created a prototype that sliced a hotdog into the shape of an octopus. His first design looked like an egg slicer. No good. Going back to the drawing board, he wanted the gadget itself to look like an octopus.

With a few design changes, his hot dog slicer looked like a friendly marine mollusk. $100,000 and thousands of test hot dogs later, he launched the Octodog on his website. It's also found in gift stores at aquariums across the country. With 7,000 units at $16.95 sold over a year and a half, not surprisingly, Ed can never look at a hot dog the same way again. Much less eat one.

Grandpa Witmer's Old Fashioned Peanut Butter Mixer™ Crank

Cranky, but Not Nuts

STAT BAR

PATENT: US #D515351-S

PRODUCT PRICE: $11.50–$13.50

STATE: Ohio

INVENTOR'S AGE: 50 years old

INVENTOR'S PROFESSION:
"My profession for the past twenty-nine years has been as director of a non-profit community service organization."

MONEY SPENT: undisclosed

MONEY MADE: undisclosed

WEB ADDRESS: witmerproducts.com

Each time he'd open his "natural" peanut butter and mix it up, he'd splash oil on his wife's pristine kitchen counters.

Peanut butter and jelly—it's a classic. It probably tops the list in comfort foods. I don't know the stats, but when it comes to lunch boxes in elementary schools, I'm sure PB&J must give tuna a run for the money.

These days, everyone is going nuts over natural peanut butter. You know, the kind without hydrogenated oil so that the peanut butter separates from the oil. I'm not sure if it tastes better than the PB we grew up on, but I'm told this natural peanut butter is healthier. Now once moms hear natural is healthier, that's what they get for their kids. But here's the problem—how do you get the peanuts to mix with the oil?

Bob Witmer has the mixed-up answer. It's called Grandpa Witmer's Old Fashioned Peanut Butter Mixer Crank. And it's only natural that Bob was the person to invent this nutty product. Bob grew up in

Orrville, Ohio—home of Smuckers. In fact, his mother and wife worked at the company for a period of time, which helps to explain Bob's passion for peanuts.

Growing up in Smuckers' backyard, Bob was always taking things apart, but not always putting them back together. In sixth and seventh grade, he tinkered with model rockets and won his school's science fair. After college, while working at a local Boys and Girls Club, he invented some wacky stuff like a spring-loaded contraption that helped kids dunk a basketball like NBA superstars.

Believe it or not, Bob's first break-through as an inventor had nothing to do with peanuts. It was a beach umbrella anchor called a sand auger (it anchors the umbrella in the sand)—an unlikely invention to come out of Ohio, as Ohio isn't known for its beaches. Actually, Bob got the idea at a family reunion at a beach house in North Carolina. The product earned him more than peanuts, selling over 100,000 units worldwide. Next came Paperguard. This handy gizmo covers the mouth of an outdoor newspaper box so your newspaper stays dry.

You can thank Bob's wife, Waunita, for giving him the impetus for inventing the peanut butter mixer. Each time he'd open his "natural" peanut butter and mix it up, he'd splash oil on his wife's pristine kitchen counters. This PB-eater needed a way to mix the peanuts and oil inside the jar.

For his first prototype, he used a wire clothes hanger attached to a mouthwash bottle cap, used as a knob. It worked, but he needed something more efficient. Bob persuaded his nine brothers and sisters to help test the various prototypes. He settled on a chrome-plated stirring rod and plastic crank.

Naturally, with Smuckers right next door, he showed the product to them. They loved it and the first year "shelled" out a coupon ad to promote the mixer. Bob has sold about 25,000 units and after two years has recouped his initial investment.

Bob took what some might call a nutty idea and gave a new twist to an American classic.

There's nothing nutty about keeping it simple

Cookie Stacker™
Definitely Not Half-baked

"Everyone who sees it wants one."

Every Christmas season, the Schreifels bake and decorate holiday treats. It's their family tradition. They spend an entire day working on their cookie creations. The Schreifels' specialties include extra-buttery sugar cookies and chocolate kisses in soft peanut butter cookies. I can almost smell them from here. Yum.

A few years ago, the Schreifels almost had a cookie calamity.

After Paul's wife, Michelle, baked cookies, Paul and their 14-year-old son, Donovan, helped decorate the cookies. (That's not the calamity.) After the cookies were decorated it was then time to box them up. Now normally they would let the cookies cool down before using sheets of wax paper to stack them. This night, however, it was getting late and the Schreifels didn't want to wait for the cookies to cool. Eyeing the wax paper, Donovan wasn't ready to see his day's work squished with one swipe. He said, "If we had little shelves to put in that container, the frosting we just put on the cookies wouldn't get smashed." They found some shoebox covers and used them as shelves, stacking them on top of each other. The lip of each cover kept the layers apart. The cookies were safe, the calamity had been avoided, and the first prototype of the Cookie Stacker had arrived.

As with most prototypes, there were problems. This cardboard version sagged in the middle. So the very next day, Paul went to a plastic thermoforming plant. Amazed? You'll probably be less amazed when I tell you that's where Paul works as a project engineer. He used his work facility to make shelves out of a flat sheet of plastic, cutting them to fit into the plastic container. He then made solid plastic dowels, which he screwed through the sheet to make legs. Paul added a foot in the middle for support. These cookies were not going to crumble.

Donovan took the Cookie Stacker to school for the annual Inventors Fair. Several teachers came up to him after the fair and said, "You really need to do something with that idea. It's great!"

Convinced they were onto something big, Paul made a third prototype out of thicker plastic. In this version, there was no sagging, even without the center support. By losing the center support, there was now room for more cookies, crafts, pizza—or whatever anyone wanted to stack. Another improvement added to this prototype was locking feet. When the shelves stack, each one locks into the one below and there's no sliding around. This invention is one tough cookie.

Paul says, "We tested it ourselves and put it to use right away. We've been using it for three years. It works perfectly for family get-togethers when we bring cookies. Our family comments on how the decorations on our cookies stay perfect. Their frosting is smashed, squished, or rubbed off on wax paper. They're jealous and ask, 'When can I get one of these?' I tell them I'm working on it."

And he is. The Schreifels are in the process of getting their patent. They hope to find a manufacturer or big company, like Rubbermaid or Tupperware that will make the Cookie Stacker part of its product line. Paul envisions a whole line of options, including different shapes, depths, and sizes. "We'd like to design the container and the shelves and sell it as a complete unit."

Paul hasn't spent any real money on the Cookie Stacker, except for the patent. "I've had conversations with our marketing and sales managers at work, and they showed some interest in producing it. But tooling would be a big expense for us. Michelle and I both go to work every day and have bills to pay.

"Besides, if we make it ourselves, we're not sure how to market it and get it out there." Their website is a start. And they made it to top twenty-four on the TV show *American Inventor,* out of over 10,000 inventors. They kept advancing until right before the show went to the last twelve that were on TV. "It was exhilarating and nerve-wracking at the same time."

"Everyone who sees it loves it and wants one." Unfortunately, they're not manufacturing the Cookie Stacker yet. "We're getting e-mails from people who saw us on the *American Inventor* show and our website. They want to buy them. We might make a few to get some out and create more buzz."

Until a hungry investor comes along, this idea is going to remain in the oven baking. The Cookie Stacker may not be done yet, but the inventors can already savor the sweet smell of success.

That's the way the cookie doesn't crumble.

Mac & Cool™ and Potty Mitts™

Cheesy Idea?

STAT BAR

PATENT: information not provided

PRODUCT PRICE: Mac & Cool $5.99;
Potty Mitts $4.59 for 12-pack

STATE: Arizona

INVENTOR'S AGE: 43

INVENTOR'S PROFESSION:
former corporate buyer

MONEY SPENT: $40,000 plus

MONEY MADE: "enough to keep
it going"

WEB ADDRESS: macandcool.com,
tudys.com and pottymitts.com

"There's nothing like having children whining to make you come up with a solution."

There's nothing like a plate piled high with macaroni and cheese. But think of all of those hours wasted as Mom blows on forkfuls of mac and cheese trying to cool it down. Junior is hungry and cranky. He grabs the fork. The mac and cheese goes in his mouth. Waaahhh—still too hot! Now he's hungry and hurting. This is comfort food?

Denise Marshall faced that scenario many times with her sons Scott, now 11, and Nate, now 8. But in 2002, she came up with the idea to fill a plastic dish with water and store it in the freezer until ready for use. Place some hot food in the dish, and the ice inside begins to melt. Soon it's cool enough for little ones to dig in. Denise called it Mac & Cool.

"There's nothing like having children whining to make you come up with a solution," says Denise. "And the bonus is that the kids get involved in getting their own dinners. A minute before the hot food is ready to serve, I tell them to get out their Mac & Cool bowls from the freezer. Once they have the food in the

bowl, they stir it and test its temperature. They get to choose the right time to eat it. It's fun for them and gives them more control."

To turn this cool dish into a hot product, Denise and her husband, Dan, dug into their savings and took Mac & Cool to market. They named their company "Tudy's" from their names—Dan and Denise. Two D's, Tudy's. 2-cute, huh? When Dan got laid off after a long time on the job, their passion for Mac & Cool didn't cool off. "After sinking $10,000 into molds and other costs, we had a choice to make—either invest the money we'd set aside for our kids' products business, or conserve it until Dan got a new job."

Cool solution, huh?

Heating up their business got the nod . . . and the funds. Tudy's started marketing Mac & Cool through the Internet and online companies that specialize in products for kids. Publicity in national magazines and guest appearances on TV kept sales hot. But Denise kept ideas simmering on the back burner. After Suzie (now 3) was born, potty training turned up the heat on Denise's creativity.

One day, at a playground birthday party, Denise saw one of the moms come back from taking her toddler to the public restroom. The look of disgust on her face told the whole story. It was a hotbed for germs and, like a zillion other toddlers, Susie would probably touch every grubby surface around the toilet.

"It suddenly occurred to me. I'd make low-cost, disposable sanitary mitts to cover a child's hands when going potty in public restrooms."

A self-proclaimed germophobe, Denise calls Potty Mitts a "wonderful addition to my arsenal of germ fighters." Made of a soft waterproof material and shaped like mittens, Potty Mitts slip on and off little hands easily. They sport a teddy bear design and come in a 12-pack for $4.59.

Launching Potty Mitts proved to be simpler than getting Mac & Cool into the marketplace. Tudy's already had the setup and potential buyers through its retailers and established website. Plus, they'd learned a lot in the two years since Mac & Cool. They sold about 10,000 units of Potty Mitts in its first year.

Denise gives their publicity effort a lot of the credit. Early on, they signed up with a firm specializing in kids' products. Says Denise, "Orca Communications got me placed on such national TV programs as *Good Morning America, Inside Edition, Dr. Phil,* and *CNN Headline News.* It was a blast showing my products on air. For a while, I got so many interviews, my kids thought seeing their mom on TV was an everyday event."

Denise also traveled to trade shows and met other women entrepreneurs in the kids' products area. At times, they'd do joint marketing, cross-selling on their websites, and getting placed in retail stores.

For both Mac & Cool and Potty Mitts, assembly and fulfillment have evolved into an all-family affair. Denise and Dan get help packaging orders from their boys, Nate and Scott. And, she may not know it yet, but Suzie's place on the production line is waiting for her when she gets a little older.

Just think—it all started by making a comfort food more comfortable.

Germbuster!

Bottle Sling™

Sling Shot

Phil Jones had his hands full. A single Vermont dad going through a divorce, on welfare, and caring for his 3-year-old son Joey and 15-month-old daughter Bridget. Phil was being pulled apart from every direction. His greatest anxiety was hearing little Bridget crying. She wanted her bottle, but dads only have two hands. "How do I feed her and hold her and use my hands, too?" he kept asking.

"Bottle Sling makes the baby number one, not the bottle."

"Think about it," he says as he remembers the juggling he did back in 1995. "When you want to hold the baby with both arms or do something while feeding her, you have to pull the bottle out of her mouth, put it down, grab a towel, and listen to a lot of crying. Wouldn't it be better to keep feeding her?"

Ironically, Phil figured out how to "grow" a third arm by looking at people who'd broken one of their own. You break your arm and put it in a sling. You need a hand feeding a baby—you make a sling.

Phil sat down and hand-stitched material together to make a sling that would go around his neck, be adjustable, and hold a bottle securely at the right angle for feeding Bridget. As he was stitching it in the kitchen, his dad walked

in and offered help—in the form of a 1957 Singer sewing machine. Between the two of them, they figured out how to use the sewing machine to make the first prototypes of his Bottle Sling. Since then, Phil has sewn thirty to thirty-five slings for friends and relatives, still using that same old Singer sewing machine.

Somewhere in all of this, Phil found the time to think about grandparents and babysitters and other caregivers who would enjoy feeding babies better if they could rely on a bottle sling. "Women who breastfeed are at an advantage—they get to hold the baby lovingly in both arms without having to hold a bottle. The rest of us are at a disadvantage. I'd like to change the way people bottle-feed their babies. Bottle

Sling makes the baby number one, not the bottle."

Joey and Bridget are now teenagers, and Phil is happily remarried to a lovely lady named Melanie. Phil has steady work as a custodian at an affordable non-profit housing project called Brattleboro Area Community Land Trust. Although he hasn't had the time and money to build a business around his invention, he did get it patented. He has a website created by his stepson Taylor that's written sort of like a newsletter. It contains articles and interviews that tout the benefits of the Bottle Sling. It even includes a video clip showing how the sling prevents a bottle from tipping and spilling.

Recently Phil had his chance at fame and fortune as a candidate on *American Inventor* television show. "I answered an ad and drove four hours to New York City for an initial audition. I was so eager. I was twenty-seventh in line. A week later, they flew me to Los Angeles for a second audition. I was so nervous. I totally choked. The judges didn't like that. I'm not a good salesman. I wasn't quick on the draw when they asked me questions. So that was the beginning and end of my television career."

For Phil, Bottle Sling is more like a hobby than a serious product. He'd be happy if a company came along and licensed his design. "I keep surfing the Web for baby companies that might be

Bottle hangs for easy feeding.

interested. After all, 4 million babies are born in the U.S. every year and 90 percent of them are being fed with a bottle within their first six months. That means the market for baby bottles is huge.

"But I'm in no hurry. I'm not going to let this idea drive me crazy. Besides, it's already given me what I really want. In the process of pursuing this venture, I've come up with a great family life.

Will having a successful, million-dollar invention change my life? Well, I don't know. What I know is that I have a happy life now."

You've got to hand it to Phil. When the going got tough, he didn't hit the bottle, he improved it.

Cozycooler™
Pass Me a Cold One

"It was one dead end after another."

Jack Hunter is a fun-loving guy who has played nickel-ante poker with the same group of guys for twenty-five years. And let's face it, when guys sit around playing cards, they need drinks—cold drinks. And in Memphis, Tennessee, especially in the summer, a cold drink can be more valuable than a pair of kings.

Jack started looking for a cooler that would hold longneck bottles. He looked everywhere and just couldn't find one. And then, as if he was dealt a winning hand, it hit him.

Jack decided he would design the ultimate cooler.

After four dozen prototypes and $20,000 from his kitty, he came up with the Cozycooler. It has two specially designed freeze packs and patented contoured walls that keep the drinks cold and cozy—without ice. And his patented two-way freeze pack works horizontally for bottles and vertically for cans.

It took this 56-year-old self-employed printing broker seven years to up the ante. "I thought I'd call a manufacturer and then a few stores to sell it. It wasn't like that. It was one dead end after another. I'd give up for a while, then I'd get another idea how to get this thing going." Jack's flexible schedule helped him pursue his cool idea.

Jack initially thought it would be a hard plastic cooler, but found that tooling costs were prohibitive—it would cost more than $50,000 to have a mold made. Instead, he made some rough prototypes out of Styrofoam and surfed the Internet to find a manufacturing company in China that made cooler bags. "I would send drawings;

they would make a sample and send it to me. I'd tweak it and they'd make it and send it back. This went on for about seven months until we finally got it right." His contact there suggested the soft foam material that he decided on. It serves as insulation while holding bottles and cans in place.

After getting the prototypes a year ago, Jack had his poker buddies and three grown daughters, Holly, Jill, and Ellen, test them out. They were game. Did this wild card have any value? While admittedly unscientific, their testing methods proved that it worked. "I knew it was a hit when I saw my buddies using the cooler over and over again—at the golf course and around town—after they'd tested the Cozycooler at our poker games. And my daughters' friends wanted to buy them. Really, it's a no-brainer. Instead of messing with ice, you've got two freeze packs that get colder than ice does. This is the cooler people grab."

Jack decided to take the gamble and go for it. He ordered a container load. He presold 1,000, so he had at least part of the costs covered. "I knew they would sell at that point, so it wasn't quite as scary."

Sales are gradually picking up, especially after a nice article in the local newspaper. Neighborhood businesses placed orders and wanted their logos on the side.

The Cozycooler is sold on a website and in local stores. It's also sold as a fundraiser. Non-profit businesses like them. He lets them keep half of the money they sell the Cozycooler for. Schools have snapped it up, as well as Executive Women International (EWI). This organization has a local chapter in Memphis and raises money for scholarships. The members started

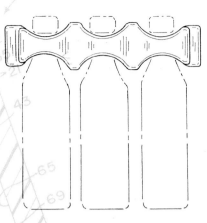

selling Cozycoolers just recently and have done well with it.

The newspaper article mentioned that EWI was using the Cozycooler as a fundraiser. As a result, a naval cadet called, interested in the Cozycooler to raise money for a group of students going into the navy. She took a dozen to her job at a hair salon and then called Jack saying, "I sold the twelve—and forty more." She's placed several orders since then. Jack explains, "Once you demonstrate this cooler to people, they want one."

Now he's trying to find people who can take this thing national. "It's hard to get the attention of Wal-Mart and other big stores unless you have a connection. I'm looking for someone who has the cash and expertise to do a marketing campaign."

Now that Jack has found a way to keep his longnecks cool, he just needs a way to heat up sales.

A cozy place for longnecks to call home

Motorized Ice Cream Cone™
A New Spin on the Old Ice Cream Cone

*"The idea for the Motorized Ice Cream
Cone came to me while I was reading
a children's book."*

An ice cream cone is a demanding food. The cone insists that you keep turning it or your hand will be covered in ice cream drippings. Most of us are willing to make the effort. We will slowly rotate our cones to avoid the drips and the mess that follows. Rick Hartman felt this chore needed to be eliminated, so he invented the Motorized Ice Cream Cone.

Toy maker and inventor Rick Hartman is a nationally known children's presenter who conducts toy-building workshops and performs at schools, festivals, and museums all around North America. He's appeared with his creations on *The Tonight Show with Jay Leno,* The Discovery Channel, and at the Smithsonian Institution's National Museum of American History. Over the past decade, he's worked with hundreds of thousands of young people at schools, festivals, and corporate family events, spreading the joys—and challenges—of inventing.

"The idea for the Motorized Ice Cream Cone came to me while I was reading a children's book, *The Very Hungry Caterpillar,* to my then-preschool-aged son." If you aren't familiar with the story, I'll fill you in. It's about a curious caterpillar that tunnels his way through a series of foods including apples, strawberries, and an ice cream cone. "Several days earlier, I'd been playing with the idea of mechanizing an everyday object for my next invention, and the moment I read this passage aloud, my heart leapt—I realized that a battery-operated, self-propelled, ice-cream twirling machine had just the

This idea can't be licked.

balance of whimsy and practicality I was looking for."

Consisting of a plastic, cone-shaped outer shell and a switch-activated rotating cup, the Motorized Ice Cream Cone serves the function of reducing chronic tongue stress and out-of-control ice cream drips. Besides those serious functions, the Motorized Ice Cream Cone also serves up fun. Ice cream lovers certainly agree; the cone has been spinning off shelves at toy and gift stores around the world.

Rick turned his Seattle-area garage into a workshop years ago. In fact, his

Motorized Ice Cream Cone isn't his first invention. About fifteen years ago, he took thumb wrestling to a whole new level when he created a mini-wrestling ring for two thumbs. It's still selling today.

Rick also came up with a gizmo for spinning ordinary yarn into cords. The colorful cords (called Crazy Cords) are then turned into friendship bracelets, key chains, headbands, and funky shoelaces. The Motorized Ice Cream Cone is just his latest kid-friendly invention.

Rick says seeing his creation sell is satisfying, but the most rewarding result of his Motorized Ice Cream Cone is seeing the reaction it gets from kids. "The Motorized Ice Cream Cone is so simple, so silly, so playful. It's an example of creativity that kids can fully relate to—an invention for the pure fun of it. It immediately inspires in children the notion that they too can come up with a great idea."

Let's face it. Working with kids can really make your head

spin. It's good to know there's a product out there that can keep up. And Rick Hartman has done the impossible; he's actually improved the ice cream cone.

Cone-gratulations, Rick.

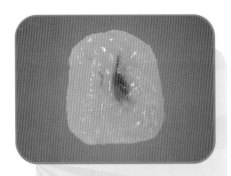

My Pet Fat^{SM}

They Say We All Start to Look Like Our Pets

STAT BAR

PATENT: information not provided

PRODUCT PRICE: $14.95/oz; $37.00/lb;
$147.00/5lb

STATE: New Jersey

INVENTOR'S AGE: 49

INVENTOR'S PROFESSION:
personal development

MONEY SPENT: about $200,000
in start-up costs

MONEY MADE: shipped 3 tons to
fifteen countries. So far it's been
pretty much break-even

WEB ADDRESS: mypetfat.com

Jay realized that fake fat could be a very real diet aid.

Let's face it. You've probably tried the Atkins diet. And heard of the South Beach diet. I bet you won't admit it, but you've probably considered trying the grapefruit and cabbage diets. So now it's time for something completely different. It could be the next dieting craze and, best of all, you don't even have to count calories or carbs. It's called My Pet Fat. Keep in mind that it's not real fat and it's certainly not a pet.

Weighing in at over 400 pounds, our inventor Jay Jacobs was losing the battle of the bulge. Things got so bad he could no longer weigh himself on an ordinary bathroom scale and, instead, ended up in the emergency room at a local hospital. Fed up with Weight Watchers, Jenny Craig, and every diet and fitness program under the sun, Jay needed a solution.

After much thought and even more donuts, Jay concluded that losing weight is not about the calories or time spent on the treadmill at the gym; losing weight is all about the mental process. So what makes people lose weight and then gain it back again? Those weighty issues got him thinking.

Then he remembered seeing an ugly silicone model of fat years before in his doctor's office. What if the next time he considered shoveling down an ice cream sundae, he could deter his craving by looking at that gross blob of fat? Jay realized that fake fat could be a very real diet aid. And then Jay had another thought: not only could the blob of plastic fat help him

Memory aid for dieters

his own. Sold exclusively through his Internet website, he packaged and shipped the fake fat from—where else—his dining room table. The media loved it and one reporter called it "outing the fat."

Now Jay is living off not the fat of the land but fat itself. The first year of business, he sold over 5,000 pounds of fat. An ounce of fat will set you back $14.95, a pound $37.00, and a five-pound slab $147.00. The more fat you buy, the more you save.

Now I'll bet you are wondering if Jay's fake fat had an impact on Jay's real fat. It sure did. Jay has lost 125 pounds. He went from 400 to 275. But Jay was not alone—the rest of the family also lost weight. Jay's wife, Kim, lost 50 pounds. His 23-year-old daughter Jennifer, lost 80 pounds and 18-year-old son, Matthew, lost 60 pounds. So the grand total for the whole Jacobs family is a loss of 315 pounds. Unlike most other gadgets, in the Pet Fat business—you want big losses.

lose weight, it could also help him gain money. Now, that's a diet!

This is more a story of reinvention than pure invention. But who cares? When it comes to fat, it's not what you have; it's what you've lost. So Jay persuaded the makers of the original fake fat (sold primarily to doctors) to give him exclusive rights to market his fake fat directly to size-challenged folks like himself. In January 2004, Jay launched My Pet Fat. His house was filled with fat, and for the first time it wasn't just

Can-o-fat

Aquariass
Johnny-Light
Plunger Holder
The Talking Toilet Paper Dispenser
Zeets Magical Seats
GasBGon
Shower Master
Shower Tan
Easy Bath Elderly Bath Apparatus
1–2 Flush
Tilt-A-Roll
Flat-D
Toothbrush Holder
Bird Diapers

5

. . . *Must Come Out*

As obsessed as we are about the food we put into our mouths, we are equally fascinated with what comes out the other end. I'll admit that there's no cable TV station devoted to bodily waste (at least not yet), and plumbers haven't reached the rock star fame of chefs, but home bathrooms have certainly been elevated to status symbols. Developers will tell you that in new-home construction, bathrooms keep getting larger and more luxurious. The room that was once never talked about is now our home's showplace. Shower jets, electronic touch faucets, and heated toilet seats are showing up in bathrooms from coast to coast. At each year's International Builders Show, the Kohler exhibit is probably the largest booth there, and it's filled with Vegas-style dancers and the latest bathroom fixtures. Really, I go every year—I mean, to the Builder Show.

As Americans embrace their bathrooms, America's garage inventors are exploring new gadgets to make the bathroom experience more "pleasant." How about a toilet that doubles as a fish tank? What about turning your shower into a tanning booth?

I even found two inventors trying to put an end to the two biggest bathroom battles: leaving the lid up and whether the toilet paper hangs over the top of the roll or along the wall. These two important issues have fractured families and have been a major source of marital strife. Could a simple gadget stop the fighting?

Bathrooms, toilets, and bodily waste—it's a part of our lives that could stand some additional enhancements. These next inventors recognize the problems, find solutions, and ultimately hope to . . . clean up.

Aquariass™

Fish 'n' Flush

STAT BAR

PATENT: information not provided

PRODUCT PRICE: $685.00

STATE: Texas and New York City

INVENTOR'S AGE: 37

INVENTOR'S PROFESSION:
metal worker, furniture maker,
and product designer

MONEY SPENT: $5,000–$8,000

MONEY MADE: breaking even

WEB ADDRESSES: aquariass.com,
aquariumtoilet.com, also
elseware.to/products/aq.htm

"It challenges what constitutes a toilet."

When we're sitting on the toilet, most of us forget about that big tank right behind us. After all, it's just filled with water and some hardware, all hidden behind porcelain walls. While we see a functional piece of plumbing, Oliver Beckert looks at that same tank and sees a piece of art.

Oliver created the Aquariass. It's a working toilet tank with a clear front and a real aquarium inside. So while you're doing your business, you have an audience of fish watching your every movement.

Why did Oliver think the world needed an Aquariass? Maybe it's because he never had a fish tank when he was growing up. Or maybe it's because he was raised in the desert of Las Vegas and needs to see water.

Whatever the reason, Oliver got the idea for his toilet fish tank while he was a student at the University of Texas in Austin studying history. It was the early '90s and Oliver had a "definite libertarian vibe that condoned and encouraged experimentation of all

kinds. For me, this meant the freedom to be creative with my living space and its objects."

Oliver's low-rent apartment lacked luxury and a toilet tank lid. He stared into his exposed toilet tank and saw a lighting opportunity.

"Water and electricity don't generally mix well, but I waterproofed a lightbulb with silicon and submerged it in the open tank, creating a soft light source and highlighting the plumbing. Beer was also very cheap and I spent a fair bit of time gazing into the swirling waters of the tank. I came to imagine fish swimming around the coppery plumbing."

Oliver went on to graduate school for industrial design at the Pratt Institute in

Brooklyn, New York. While there, he found himself working at a job that let his creative side atrophy, but it did provide him with a space for the after-hours fabrication of his visions. Oliver and four other young designers formed the Elseware Design Group. "Our approach to design was conceptual with a reverse marketing twist—build it and see if they come."

Oliver's first project was that fish-tank toilet he imagined while drunk on beer and staring into his bulb-lit tank back in Austin. He created a fish tank that mounts onto a toilet bowl tank. Oliver quickly realized you can't put the fish into the actual tank or you would lose them all each time you flushed. That would be both sad and expensive. To prevent that, Oliver placed a translucent tank inside the larger fish tank. When you flush, you see the water go down in the translucent tank, but the fish never feel a thing.

The tank mounts on an American Standard Cadet toilet bowl just like a regular tank. The fish tank holds about six gallons and the flushing water can be adjusted to about two gallons (a bit more than the federal standard of 1.6 gallons).

The Aquariass was introduced to the art world during a NYC show the Elseware Design Group called, appropriately, The Watershed Show. It featured

Tanks, for turning a toilet into art.

all bathroom products. Hundreds walked through the show and most were crazy about this unique toilet. The *New York Times* ran an image of it in its Home section that week. Oliver knew he had a royal flush.

The Aquariass went through many redesigns over the next four years; the first one was installed in 2003 in a child's bathroom in a Boston suburban home.

There are still only a handful of Aquariasses installed throughout the United States. This has not kept the press from showing continued interest. Its image has appeared in over a dozen diverse publications, from French *Playboy* to the Japanese *Wall Street Journal*, and it's been filmed for TV four times. Internet interest has also been intense. On any given day, dozens of images of the Aquariass can be found on blogs throughout the world. "It challenges what constitutes a toilet and what is possible in a bathroom. It appeals to a wide variety of people and will no doubt continue to raise eyebrows and smiles."

The tank costs $685 and comes complete with all the flushing hardware except the bowl, which comes in a number of colors and styles starting at around $60.

Oliver does not want to mass-produce this toilet because he wants to keep its quality high. He's hired a company in California to help him in the fabrication process. It takes about a day to make one tank.

His family worries because he hasn't seen any real financial gains from the Aquariass. Oliver's friends support him, but also hope it will not be his only contribution to the world.

By adding fish to the mix, Oliver Beckert has added life and beauty to the bathroom and changed the way we view toilets. Of course, no one is asking the fish what they think of their new view.

Johnny-Light™

Make Pee-ce, Not War

STAT BAR

PATENT: information not provided

PRODUCT PRICE: $14.00
(includes shipping)

STATE: Texas

INVENTOR'S AGE: 53

INVENTOR'S PROFESSION:
electrical engineer

MONEY SPENT: $100,000

MONEY MADE: "lots"

WEB ADDRESS: johnny-light.com

*"'Quit' is not in my vocabulary.
But 'survival' is."*

As a gender, men are famous for leaving the lid up and needing the light on. Both drive our spouses crazy.

For Bill Bradford, the bathroom light was a big issue. It was so bright, it would wake him up and he'd have a tough time getting back to sleep. The light also bothered his wife, Barbara. Bill knew it was a sore issue when Barbara would say to him, "Why do you have to turn that light on? I never have to—why do you?" Bill explains to women everywhere what every man already knows: "A man, by nature, does one of three things to take care of his business in the middle of the night: (1) He turns on the bright light, waking everyone up; (2) He walks until he feels cold porcelain on his shins and aims until he hears water; or (3) He sits down. Now, here in Texas, we don't like to talk about the third option—it's just not manly." There, he said it.

Being the gentleman Bill is, if the light bothered his lady, he was going to take care of business. He sat down, thought about the problem, and was Johnny-on-the-spot. He made just one prototype of his idea and called it the Johnny-Light. It's a battery-powered device that lights the toilet bowl with an easy-on-the-eyes green glow when the seat is up and turns off when the seat is down.

"I took a toilet seat, mounted the Johnny Light on a piece of wood, and that was my calling card. I'd walk in with this toilet seat under my arm, saying, 'Have I got a deal for you!'" Bill took his "commode-ity," as he calls it, to trade shows, displaying it in an outhouse painted like his packaging—night blue with the moon and stars. A black curtain at the front provided darkness for the Johnny-Light to strut its stuff on the toilet seat inside. Bill let people go in, lift the toilet seat, and see for themselves. "That's where you really see which homes have the problem. The men scurry by, not wanting to stop. The women slide on in and take a peek."

Other uses seemed to come out of the blue—or green. Besides allowing everyone to sleep at night, the Johnny-Light helps keep the rim clean because it serves as a directional beacon. As one testimonial writer says, he no longer has to "aim and shoot and hope for the best." If a woman goes into a dark bathroom and sees the soft green light, she won't reach or just automatically sit. And the green light encourages everyone to "close up shop." The most surprising bonus use is for potty training. One mother wrote to Bill, calling his invention a Johnny-*De*light because it gave her son a new reason to say, "Green means go!" Potty training her son was a breeze.

"After I'd had the patent for a year, I had enough money to order 10,000

Green means go.

units. Paying stay-at-home moms to set up Johnny-Lights, it became a cottage industry. Then I would do the packaging myself. I had to take baby steps." About a year and $100,000 later, sales are great! "I'm in a major home improvement store (Lowe's) nationwide, and Internet sales are going like gangbusters. And I've sold internationally. So that tells me the problem exists in other countries, not just in the U.S."

John Thompson, a business professor at Texas Christian University, has had Bill speak to his classes about his success marketing Johnny-Light. He was intrigued with how Bill was able to obtain over $1 million in publicity without spending a dime. The Johnny-Light has been featured on NBC's *Today Show, Live with Regis and Kathie Lee, The Leeza Gibbons Show,* and *The Howie Mandel Show,* and has been written up in *Men's Health* magazine, *Glamour, Modern Bride,* and the *Dallas Business Journal.* Bill says, "Someone would write about it and from that writing, someone else would contact me. It snowballed. I haven't paid for publicity yet."

The Johnny-Light even went to the Oscars in 2006. Bill's wife and daughter got photos taken with celebrities. The Johnny-Light was a phenomenal hit there.

His friends can't believe he stayed with it during those hard, early days. Bill credits his persistence to the way he was raised. He grew up in foster homes, then the New Mexico Boys Ranch, and entered manhood in the Marines. "'Quit' is not in my vocabulary. But 'survival' is." These days, Bill is more than surviving—he's thriving. And as his mission statement says, he is "lighting the way to family harmony."

NEW!

Johnny-Light™

Lighting the Way To Family Harmony

★ Lights toilet bowl at night and goes off when the seat goes down.
★ Ends the bathroom battle of the sexes.
★ Reminds us to put the seat down.
★ Eliminates annoying "fall-ins."
★ Aids children's toilet training.

★ Easy installation

Plunger Holder™
Flushed with Excitement

STAT BAR

PATENT: US #6729470-B2

PRODUCT PRICE: not developed yet

STATE: Illinois

INVENTOR'S AGE: 43

INVENTOR'S PROFESSION:
district manager for Chicago
Transit Authority

MONEY SPENT: $18,000

MONEY MADE: $0

WEB ADDRESS: none

"With a decorative design, attractive colors, and a place for candles on top, it's an attractive piece of furniture."

You're at a dinner party in a friend's home. At some point during the evening, you excuse yourself, and get up to use the facilities. When you've finished doing your business, you flush. That's when it happens. The toilet becomes clogged. You flush again. The water continues to rise, as does your blood pressure. You panic. You start to look for a plunger. You open the cabinet doors under the sink and anywhere else that might store a plunger. No luck. Your embarrassment mounts.

Elizabeth Watlington knows that experience and wants to end it for all of us. She believes every bathroom should have a plunger "at the ready." It's just as important as placing a toilet brush strategically near the toilet.

But let's face facts. Plungers are just plain ugly and don't complement any bathroom décor. That's why Elizabeth came out with an attractive, space-saving Plunger Holder, complete with a drip tray for easy cleaning, air holes for ventilation, and space to hold candles on top.

This inventor sees the Plunger Holder as much more than an embarrassment prevention tool. "With a decorative design, attractive colors, and a place for candles on top, it's an attractive piece of furniture."

Before plunging into making a prototype, Elizabeth floated the patent drawings by Rubbermaid, well known for making kitchen and bathroom accessories. But they couldn't come to terms. "I would have had to sign my life away." After flushing away thousands of dollars with a consulting company that didn't come through, she worked with an Illinois law firm to

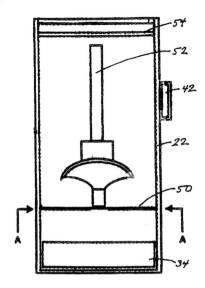

the district night manager for the Chicago Transit System and also being a full-time B.A. student at Governors State University. The Plunger Holder often comes in third. Still, she won't let her Plunger Holder dreams get stopped up. "I understand how time-consuming the process can be. I need to be cautious and patient, but I'm determined to follow through and someday have licensing fees and royalties to pass on to my kids."

A single parent, Elizabeth has raised a son, 24-year-old Don-Tae, and a daughter, 18-year-old Amber. She also wants to leave money to her 4-year-old grandson, also named Don-Tae. Her

motivation has been fueled by a resolution she made after her oldest son, Demorie, was murdered on the streets of Chicago in 1998. "This experience made me stronger. I've had to stay focused and stay strong."

Elizabeth has proven she can rely on patience and persistence—from starting as a part-time bus operator to getting multiple promotions and earning her degree. She never gave up on herself and won't give up on the Plunger Holder. At the moment, the process may be a little clogged. But there's no doubt that Elizabeth will get things moving again soon.

get her patent approved in 2002. She also got as far as second call on the *American Inventor* TV show—and was relieved it didn't go further. "They expect you to be available over a period of four months. I couldn't possibly do that with my job, school, and family commitments."

She's now researching funding options to take the Plunger Holder to the next level. Exactly how her Plunger Holder will look—high-fashion ceramic or budget-minded plastic—depends on what happens next. In the meantime, she's writing letters to companies in the bath accessories business to get them interested in her idea and design.

Keep in mind that Elizabeth is juggling the Plunger Holder with her job as

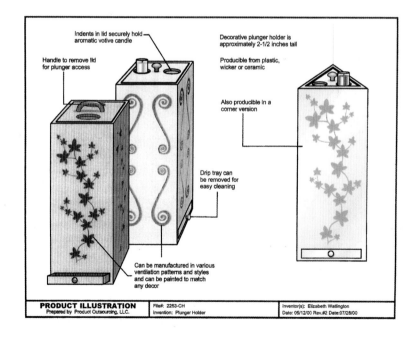

Indents in lid securely hold aromatic votive candle

Handle to remove lid for plunger access

Decorative plunger holder is approximately 2-1/2 inches tall

Producible from plastic, wicker or ceramic

Also producible in a corner version

Drip tray can be removed for easy cleaning

Can be manufactured in various ventilation patterns and styles and can be painted to match any decor

PRODUCT ILLUSTRATION
Prepared by Product Outsourcing, LLC.

File#: 2253-CH
Invention: Plunger Holder

Inventor(s): Elizabeth Watlington
Date: 05/12/00 Rev.#2 Date:07/28/00

The Talking Toilet Paper Dispenser™
Who Said That?

STAT BAR

PATENT: US #6628789-B1

PRODUCT PRICE: $19.95

STATE: California

INVENTOR'S AGE: 64 now, 61 when patent issued

INVENTOR'S PROFESSION: sales

MONEY SPENT: $15,000

MONEY MADE: "I put product into my company with others so that is impossible to say. However, revenues would be about $350,000 paid to us. Our production costs would be about $250,000."

WEB ADDRESS: talkingtp.com

His friends thought he was nuts . . .

Consider the following scenario. You're sitting in the bathroom at your friend's house. You've finished doing your business and you reach for the toilet paper. As you pull the roll, you hear your friend's voice tell you: "Hey buddy, don't use up all the toilet paper!" Huh, where did that come from? Am I on MTV's *Punk'd*? Are they bringing back *Candid Camera*?

No, fortunately, you're not on some new bathroom-based reality show. Though frankly, that reality show is probably in production somewhere. In this circumstance that disembodied voice came from the bowels of a Talking Toilet Paper Dispenser, and you can thank inventor Russ Colby for your disturbing message.

Russ wasn't always into bathroom scare tactics. He got his start at Georgetown Law School and went into securities and compliance for a big brokerage house in New York City. Bored with law, he went into sales, selling everything from financial printing services to medical products.

Russ got into his own business selling "panic button" systems. You know, transmitter pendants you wear that when pressed call a para-

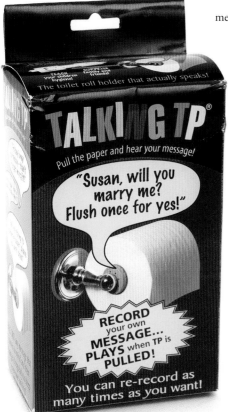

medic. Remember, "I've fallen and I can't get up?"

One day, Russ joked with a friend: "Imagine if your TP could talk." I can't, but Russ did, so he went to work. His friends thought he was nuts, but that only encouraged him. Russ felt he was "flush" with a great idea. He got his inspiration and motivation from his father, who was sort of an inventor. His father owned a toy store and came up with the idea of "invisible fish." He put an empty fish tank in the window with a sign stating "Invisible Fish." Kids came by and fed the invisible fish with invisible fish food. How's that for clever?—
or is it fraud? That's a question for a future book.

Anyway, six prototypes later, this 61-year-old came up with his product and found a distributor to take it to market. He's flushed a million and a half dollars in sales out of consumers (surprisingly, mostly women) and a kids' version is on its way.

Talk is cheap, but a Talking Toilet Paper Dispenser is $19.95. So heads up (or bottoms, if you will) next time you reach for that TP. Good going, Russ.

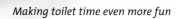

Making toilet time even more fun

Zeets™ Magical Seats

Best Seat in the House

STAT BAR

PATENT: "yes"

PRODUCT PRICE: $12.99 for a box of 5 Zeets; $95.00 for 10 boxes of 5

STATE: California

INVENTOR'S AGE: 52

INVENTOR'S PROFESSION: lawyer, business consultant

MONEY SPENT: $20,000–$30,000

MONEY MADE: from a few thousand units sold

WEB ADDRESS: zeets.com

"Most inventors think people will see their products' value instantly and run to buy them, but it doesn't work that way."

When Mark's daughter Devon was a toddler, using a public toilet was a major ordeal. His wife, Erica, would hold Devon over the toilet so the toddler wouldn't fall in or have to sit down on germ-filled, disgusting public toilet seats. They both hated public potty time. Mark recognized the situation stank and decided something had to be done.

"Manufacturers have been creating child-sized toilet seats with fun designs for sixty years. Why not portable toilet covers for kids to use when they're not at home?"

Mark knew the idea wasn't new. His sister had given him a plastic folding seat for Devon after her son outgrew it. "But it wasn't disposable. After one use, you'd have to fold it up, bring it home, and wash it. We'd be carrying all those germs home with us." Nasty, huh? So Mark created Zeets Magical Seats, a disposable folding cardboard seat designed with a small opening for a child's behind. Use it once, then throw it away.

Next came marketing. Mark came up with a character named Zook, and Erica, a children's author, wrote a story about an alien named Zook, from the planet Veek, and Zeet who's "like a pet, a robot, and a best friend all rolled into one." Zeet acts as a protective shield that flies Zook all around the galaxies. In the story featured on the seat's packaging, Zook leaves Zeet with a child named Devon while he goes on a special mission. While Zook is away, kids can take Zeet on their own special missions—into public bathrooms.

Zeets Seats come packaged five in a box for $12.99 and Zeet-lovers can buy a case of ten boxes for $95. Measuring approximately 7 inches square and weighing 3.4 ounces, a Zeets seat cover can support the weight of a child up to forty pounds. "My daughter Devon still uses it, even though she weighs well over that amount. She knows not to lean back on it and rip the cardboard."

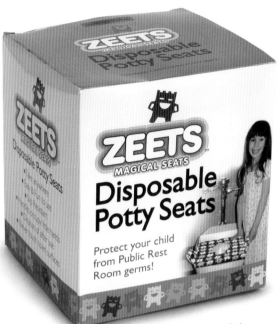

A 52-year-old lawyer and business consultant, Mark worked out the kinks in the early stages with several designers. He found it easy to apply for a provisional patent and got through all the technicalities for the utility patent with the help of a patent attorney. "This can be a frustrating process, especially because I'm familiar with legal procedures. The people who approve patents know engineering well, but they're forced to interpret patent laws when they're really laymen."

Publicizing and marketing Zeets has been tough. "Most inventors think people will see their products' value instantly and run to buy them, but it doesn't work that way. Even if you do a great job of promoting, a product may not catch fire." Mark admits he's not a keen marketer. He's approached at least fifty baby products retailers to carry Zeets, with marginal success.

In the retail world, volume counts most. "I only have one product and not a line of them. The big retailers won't buy unless you've proven there's a craze for it . . . like a hula-hoop. Otherwise, it takes a huge budget to promote any product." Still, he's sold several thousand in three years through his website and a handful of retailers.

When Zeets first came out, it hit what would seem like the publicity mother lode; it was featured in *USA Today*. But that coveted exposure didn't yield results. "We didn't get a single order from that. It also got mentioned in *Daily Candy*, an e-mail newsletter that brings attention to innovative things. We got 60,000 hits from that, so I bought a few ads on *Daily Candy* without reaping nearly that volume of responses. Now that Zeets isn't a new product anymore, we may have missed the window of opportunity that such publicity provides—unless something dramatic happens."

He has come up with one publicity opportunity that could prove fruitful. *InStyle* magazine looks for all sorts of ways to put news about celebrities between their pages. "If a celebrity started using Zeets, then I'd have something to talk about for *InStyle*'s audience." Know any germophobic celebrities with kids who want to brag about Zeets?

Making public toilets a little more user-friendly

GasBGon®

Got Wind?

"GasBGon levels the playing field between partygoers and party poopers."

Let's be honest. Everyone passes gas. It's an ugly reality. Most of us just try to ignore it . . . and hope the people close to us don't mention it. As quickly as the gas passes, we put it *behind* us and move on. But unlike the rest of us, Sharron and Jim Huza faced this derrière dilemma and decided to put an *end* to it.

Sharron was enrolled in an Internet marketing course. For her final project, she had to come up with a mock product and website. She played around with a few ideas, then put the project on the back burner.

That weekend, Sharron and Jim hosted a "pig pickin'" party at their house. The weather was not cooperating, so the party moved indoors. The side dishes consisted of baked beans and broccoli salad, as well as cabbage casseroles, and to wash down all this fine food, lots and lots of beer. Can you see where this story is going?

After a few hours, the Huza home started to take on a less than pleasant odor. Sharron told Jim to open the windows or do something about the awful stench. It was an "aha" moment. Sharron's marketing project would involve a product aimed at gas control. So Sharron and Jim put their heads together to come up with the GasBGon Flatulence Filter Seat Cushion.

Per Jim, "I'm an application engineer involved with providing corrosion and odor-control solutions in the pulp and paper industry. There, the host contaminants that cause problems are sulphide gases, the same gas we smell when we pass. Sharron submitted her product idea and received 125 out of 100." What started as a mock marketing project exploded into a real product.

Here's how GasBGon works. "We have replaceable tandem filters consisting of an acoustical foam filter to muffle the sound and diffuse the outburst, followed by a proprietary blend of activated carbon with enhancers to absorb the odor, which has the equivalent surface area to that of a football field."

The Huzas wanted to manufacture GasBGon in the U.S. as much as possible. North Carolina has a lot of people who sew professionally, so making the cushion covers was easy. To assemble the cushions' components, they stumbled on the East Carolina Vocational Center, which employs disabled adults.

Because of the subject matter, Sharron and Jim decided to introduce GasBGon using a humorous approach. They created cushion patterns with different titles. The leopard print is called Silent Butt Deadly; a print with musical instruments is the Musical Solo; soccer print is the Back Pass; golf print is the Bunker Shot; the basketball print is called Real Foul. For those less inclined to advertise, the Model-T is basic black.

"One of our newest incorporates university and college color schemes, perfect for a tailgating party. GasBGon levels the playing field between party-goers and party poopers."

Sharron and Jim tell me they had no trouble marketing GasBGon, never paying for advertising. They sent out media releases and learned that news organizations love flatulence. Hundreds of radio stations, newspapers, and television newscasts responded. At one point, they were doing thirty interviews every day for two months.

GasBGon had become a media darling. And for some unknown reason, when they did CNBC's *Squawk Box,* they received their biggest boost in sales. Maybe that says something about the viewers of *Squawk Box?*

All of this attention was great for sales, but created a stink back home. The Huza kids weren't proud of their parents' endeavor. "When Stephanie was 12 and Jamie was 8, they would tell us they didn't want to go school. I remember them saying, 'We're going to tell everyone that we're adopted. This is just so embarrassing.'"

Not to worry. The Huza kids got behind GasBGon when a local TV news station decided to shoot a commercial about the product. The station wanted to have the kids in the commercial and the Huza kids wanted to be on TV. EmbarassmentBGon.

Sharron and Jim named their company Dairiair® and have a great attitude and an even better sense of humor about their product, but they've learned that passing gas can be serious problem. For people with diabetes, irritable bowel syndrome, and MS, it's no laughing matter.

Silent Butt and others

To meet this need, the Huzas developed an undergarment called the GasMedic™ Nether Garment. It's made of a new lightweight odor filter that can be washed and used over and over.

Once again, they had no difficulty getting the media to cover this product. Sales hit a new high when Lisa Rinna endorsed the product on her *Soap Talk* show. Sharron and Jim say, "We've received hundreds of e-mails, letters and calls from people who said this product gave them back their lives."

When Sharron and Jim first went into the flatulence business, they didn't get the support of family and friends. "Everybody said, 'You've got to be kidding.'" But today, the Huzas get the last laugh. "What everyone thought was so funny ultimately became so successful."

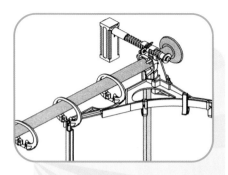

Shower Master™

All Wet?

STAT BAR

PATENT: information not provided

PRODUCT PRICE: will be between $30 and $40

STATE: Pennsylvania

INVENTOR'S AGE: 70

INVENTOR'S PROFESSION: engineer, retired federal government employee

MONEY SPENT: $400,000

MONEY MADE: none

WEB ADDRESS: none

Tony has manufactured about 10,000 Shower Master units. Sadly, most are still sitting in his garage.

The whole purpose behind a shower curtain is to keep the water on the person in the shower and off the floor, right? But we've all been acquainted with shower curtains that just don't do their job. You step out of the bathtub shower only to find a very wet floor. It can really get you really steamed.

That wet-floor feeling bothered Tony Ruggiero when it happened in the tub-shower of his Philadelphia home. Sure, wet floors are annoying, but Tony recognized a wet floor can become a serious problem. It can lead to long-term problems such as mold growth, mildew, wood rot, and even costly damage to walls, floors, and ceilings.

It was when Tony became a landlord that wet bathroom floors began to bother him even more. "Most renters won't make a conscious effort to clip or seal their shower curtains to the tub walls. They don't really care—but as a landlord, I do care." He had to find a way to make sealing a standard shower curtain a no-fail proposition.

So Tony, a handyman by avocation and an engineer by profession, created the Shower Master Spray Control System. It's a device that seals a standard shower curtain to both corners of the bathtub. You don't even need to make a conscious effort to use it—just step inside the tub and the device automatically seals the curtain. It has no clips, Velcro, suction cups, or anything to fasten or unfasten—it simply closes the curtains. The Shower Master is even hidden from view when a decorative outer shower curtain is used.

CURTAIN RING
(12)

LOCKING PIN (2)

ROLLER
(2)

Being a landlord has showered Tony with lots of testing opportunities for the Shower Master. He has placed Shower Master in all of his rental apartments. He sleeps easy knowing his bathroom floors aren't all wet.

Tony has manufactured about 10,000 Shower Master units. Sadly, most are still sitting in his garage. He's learned a tough lesson—people are lazy and most are not mechanical. Tony's original Shower Master design required the user to put pieces of it together. People liked the idea of keeping their bathroom floor dry, but setting up Shower Master intimidated most consumers. So they'd send it back. "I call this my invention that everyone can use and no one wants."

So far, Tony has sunk $400,000 into the Shower Master, but he doesn't feel it's been money down the drain. Combined with his wife Karen's pension

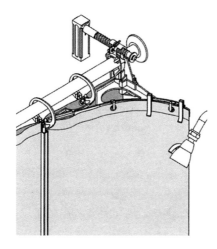

and his own pension from thirty-eight years with the federal government, making money with his Shower Master hasn't been a big incentive for this 70-year-old.

Then you add the real estate factor.

"You see, I knew it would take a lot of money to bring an invention into

the world, so back in the early 1990s, I bought some rental properties in a slow area of Philadelphia. Although my invention has fizzled, the real estate has skyrocketed." With proceeds from the twenty-four rental units Tony owns and manages, he's been able to recover the money he's spent to develop the Shower Master.

As for Tony's kids, they consider the Shower Master "just one more thing Dad's involved in," so they don't pay much attention.

Tony foresees a huge market for the Shower Master among hoteliers, homeowners, and landlords. After all, it's a simple, easy way to prevent damage to bathroom floors. But Tony says the Shower Master has one barrier—the price. To make it profitable, he would have to sell the Shower Master at a price between $30 and $40.

Tony has a new-and-improved design ready to manufacture. He still needs to put together the right combination of mold builders, manufacturers, and distributors. But he says he is unsure of his next step. "I think, 'This is too good to let slip away.' But then I wonder, 'Is it really worth all the effort it would take?'"

Does that mean it is "curtains" for the Shower Master? I doubt it. Tony Ruggiero knows the world needs his Shower Master and he'll find a way to make a splash.

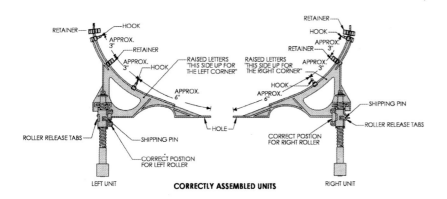

CORRECTLY ASSEMBLED UNITS

LEFT UNIT   RIGHT UNIT

Shower Tan™

Burn, Baby, Burn

STAT BAR

PATENT: pending

PRODUCT PRICE: $250–$300

STATE: Florida

INVENTORS' AGES: Julie, 31; Rick, 32

INVENTORS' PROFESSIONS:
Julie, physical therapist;
Rick, mechanical engineer

MONEY SPENT: $30,000

MONEY MADE: haven't taken
anything out of the business;
sold $18,000's worth

WEB ADDRESS: showertan.com

"A lot of people think we're crazy."

Living in Florida, Julie Buswell has always kept up her tan—at a salon. In 2003, she tried UV-free spray tanning. It cost $30 a session. She told her husband, Rick, about it and they talked about opening a UV-free tanning salon in Tampa, but learned that spray tanning booths cost $30,000 each.

Julie said, "Rick, you're an engineer. Couldn't you make one for a lot less?" That's all Rick needed to hear. Now he just needed to come up with a bright idea. Julie wondered, "It would be really great to make something people could use at home. What if we had something that could spray in the shower?" and Rick was pumped.

First he bought a pump, but couldn't get it to work. Next, he tried carbon dioxide (CO_2) as a high-pressure power source. The concentrated tanning solution would come out with the water, then water pressure would do the work. With tubing and nozzles attached to suction cups, Rick's device hung on the shower, sort of like a shower radio. The first prototype worked, using a CO_2 gas cartridge. Julie eagerly stepped in and be the first to try it.

An even tan later, they decided a high-pressure gas cartridge was the way to go. To perfect the technique, they tried different amounts of tanning solution and kept changing the number of nozzles. As the guinea pig, Julie kept getting darker and darker.

"The first prototype was really heavy and had user-friendly issues. For example, our shower has a glass door and tile walls—suction cups don't work well on bumpy tile. So we came up with the idea of having something that you set down on the floor, and it stands up and sprays you. Here we are, three years later, with the Shower Tan."

Rick and Julie found a small engineering company to help them improve the design. One question was pressure. The device needed to hold 1,000 psi when you release the CO_2 cartridge.

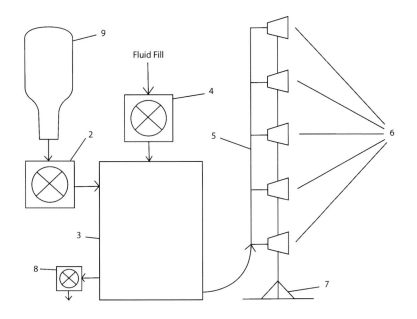

Fluid Fill

They made two prototypes. Remembers Rick, "We were spending money, but still felt encouraged by the product's potential."

"A lot of people think we're crazy," says Julie, "to spend the amount of money we've spent. They ask, 'How much did you spend? And you're going to spend more?' But they've always been interested and want to try it."

Rick says his brother-in-law, John, lives in Tampa, too. "John always says, 'I think this will be your learning business. You'll make it someday, but I don't think with this business. I wouldn't quit your day job over it.' But people who know about sunless tanning think it's a great idea. Our friends and family don't tan, so they just see the dollars going down the drain. We're encouraged by the people who write us notes that say it's a great idea."

The Buswells ordered their first set of parts and built ten Shower Tan units to see if they could sell them. Next, it was time to buy an Internet domain name. Talk about burn. Rick had checked on showertan.com and found he could buy it for $20. But six months later, when he was ready, a squatting company had swooped in and said, "Tell us your interest and we'll send you a bid." The bid came in at $395! Six months later, Rick and Julie decided Shower Tan was indeed the name to use. So they con-

tacted the owners—and found the going price was $795. "It was outrageous—I wrote and was told, 'Sorry, with the demand for that name . . .' We were probably the only ones demanding it, since we'd con-tacted them twice. So we negotiated, and bought it for $500."

The Buswells had more success with their marketing. Rick explains, "We had the Shower Tan on eBay and had been getting one or two visits a day. I noticed we were getting traffic from Sunless.com, so I went to the site. Someone had written, 'Hey, has someone seen this Shower Tan from eBay?'"

Rick made sure that happened. He contacted Vicky Mayhew, the product tester for Sunless.com.

She was willing to test the unit and post her impressions on the site. After reading her first post that said, "I just sprayed myself. I'll let you know how it comes out . . . " Julie and Rick held their breath. Would Vicky love it or pan it? Her review came out and Rick tells me, "She loved it! Tons of people responded, saying, 'Mystic Tan should go this route.' . . . A month later, those first ten were sold out." They quickly ordered parts for forty more. You can read Vicky's review on the Testimonials link on Shower Tan's website. More on Mystic Tan later.

"The business has really picked up this year. In six months, we've already sold as many units as we sold all of last year. It's a lot of work. Neither one of us has quit our job. Julie works with handicapped children at a school and I'm a manager for a bearing manufacturer, in a sales/engineer/ account management position. I just stay up late a lot. We're at the point where I'm getting maxed out. I have to keep up with the assembly, billing, and shipping, as well as answering questions that come in."

Rick and Julie are now talking to people about writing a business plan for a full-scale launch. They also need a more user-friendly package design before they're ready for a store shelf. Then they'd like to sell Shower Tan on QVC. They're cautiously optimistic. "We've had lots of things that got us excited, but didn't lead to anything," says Rick.

They tried out for the *American Inventor* show and two weeks later got a call saying, "You've been selected for the next round."

"It was really exciting!" Rick says. "But two weeks after that, they called us back, 'Sorry, there's been a change in the show's direction and you've been taken off the list.' We still don't know what happened."

They also got a call from the founder of Mystic Tan, the company that marketed the first spray-tanning booth. Rick had given a brochure to the marketing guy at a show, trying to find somebody to license it. "All along, we've felt this is bigger than what we can do ourselves. If we can just get a piece of it, and let another company do all the work, that would be great."

It seemed to the Buswells that Mystic Tan might be interested in licensing the Shower Tan. "He asked about our patent and how the Shower Tan worked. He wanted a unit to test and try out. I asked if he'd sign a confidentiality agreement. He said, 'No, we don't do that. What's the point—it's on the market already anyway.'

"A month later, we got an order from him, using his name. That was another high. We thought, 'We're about to hit payday!' Nothing. I called him—he's too busy. Then he called to ask a question about Shower Tan: Is there a way to stop it? (You see, the Mystic Tan device sprays, then stops so you can turn, then sprays your back.) We'll just have to wait and see."

Who knows? As they get more attention and sales, the Buswells just might find their place in the sun—without getting burned.

Bathroom-bound Shower Tan prototype in garage

Easy Bath Elderly Bath Apparatus™

Cleaning Up?

At 71 years old, mechanical engineer Abram Fainberg knows that hopping in and out of the bath gets tougher as we get older. For people who are frail, elderly, or disabled, it presents a serious obstacle to independence and enjoyment of everyday life.

A Russian immigrant, Abram has tremendous sympathy for those, including friends in his Rhode Island neighborhood, "who have serious trouble taking a bath. It's a big tragedy when disabled, infirm people can't enter the tub and bathe themselves. In many, many cases, the people simply live with pain, tears, and deep mortification."

Abram saw the problem and engineered a serious solution—Easy Bath. Think car wash. Crafted in a rented machine shop, the prototype was tested in his own apartment, but has yet to be made or marketed.

With Easy Bath, the person—like the car—remains stationary, sitting in a standard-size chair extending over the lip of the bathtub with feet resting on a tray outside. The bather, surrounded by a shower curtain on a U-shaped rail, angles the whole chair apparatus back, causing the footrest tray to extend. Then a movable showerhead on a hose is used to clean the legs, arms, and other body parts within reach. A sprinkler built into the chair itself washes the bather's bottom.

Another part of the Easy Bath apparatus props up against the back wall of the tub and holds a washing belt that goes up and down a pole. The belt soaps, scrubs, and massages the bather's back. A pump circulates water from the tub up through the back-scrubbing apparatus so the bather doesn't even have to move.

Abram has patents on both parts of this Easy Bath invention. Now he's hoping an interested sponsor will step forward. His goal: to make life easier for millions of people, and eventually his as well. With the aging population, Abram Fainberg could really hit pay dirt.

STAT BAR

PATENT: US #6817045 and 6912740

PRODUCT PRICE: not developed yet

STATE: Rhode Island

INVENTOR'S AGE: 71

INVENTOR'S PROFESSION: retired mechanical engineer with Ph.D.

MONEY SPENT: not much

MONEY MADE: none

WEB ADDRESS: none

"It's a big tragedy when disabled, infirm people can't enter the tub and bathe themselves."

*"You get your money back in four
months! What other product pays
for itself like that?"*

1–2 Flush

Down the Drain

In the 1980s, Cabbage Patch Dolls and *The Cosby Show* became popular. So did low-flow toilets. Here's a little toilet history lesson.

Low-flow toilets were part of a push to care for our environment and save water. A low-flow toilet meant switching from a seven-gallon tank to a three-gallon tank, but still using the same type of bowl. The toilets worked fine with the large volume of water they were designed to handle. But with a reduction from seven to three gallons, water wasn't flowing through with the same force. Often, one flush didn't do the job, so a second flush was needed. Since you have to wait for the tank to fill before a second flush will work, it's annoying, not to mention the fact that you don't save much water when you flush twice.

Working as a handyman in the late 1980s, Edward Heath heard his clients complain about the transition to low-flow toilets. Flushed with excitement, Edward invented an easy solution: the 1–2 Flush. It's an easy-to-install handle and lift-arm assembly for the toilet, converting a gravity type, single-flush toilet

(which most of us have) into a dual-flush toilet. Dual-flush toilet means you have a choice—a full or a half flush.
If you decide on the half flush, you use only half of the water of a full flush.
If you pick a full flush, you flush all the water down the drain.

The first type of flush is the partial flush for liquid waste (think number one). The second type of flush gives you the full tank to whisk away solid waste (think number two). You choose your flush depending on whether you put number one or number two in the bowl.

Here's how it works: Once you've installed the 1–2 Flush, the handle no longer goes all the way down—only about a third of what it used to. So for the first flush, you hold it for a couple of seconds, get a little swirl going in the bowl, then release the handle. That's for a light, liquid flush. For a full flush, you push the handle in about a quarter of an inch while pushing it down. Then the handle moves in its normal, full rotation. The flapper goes all the way up, allowing all the water in the tank to flush away the solid waste. It automatically resets itself for the first (the water-saving) flush, even if the last flush was the full flush.

"My invention will work on any toilet with a flapper, not one with an air-assist power flusher. Those are turbo toilets. But there are lots of toilets out there with flappers—400 million in the U.S. alone." The 1–2 Flush is patented in the U.S., Canada, the U.K., and China, and will soon be in Hong Kong and Taiwan. And Edward is confident he can flush out the competition. "Other retro-fit products are out there, but require a lot of work. Mine is simply a handle replacement. You tighten the nut down and adjust the chain once—that's it. My product has a five-year warranty, the longest in the industry. And a decorating feature sets mine apart from the bronze or chrome handles you can get at Home Depot. I can powder-coat mine, so I can customize the 1–2 Flush to any color."

News about the 1–2 Flush has been in a local Connecticut newspaper. And it's received favorable comments when it has been demonstrated at fairs and invention shows. "I built a crystal clear tank out of Lexan. I've got the only bulletproof toilet tank in the world! I made a two-gallon tank, filled it with blue water, and let people try it for themselves. They think it's just some sleight of hand when I do it."

Edward is trying to get his 1–2 Flush into hotels and apartment buildings. "There are 72,000 toilets in one development of New York Co-op City alone!" He figures that a hotel would save 300,000 cubic feet or 210,000 gallons of water a year, with an annual dollar savings of $12,000. Average families of four would save 17,520 gallons a year. If they're on city water and sewer, their annual money savings would average $93. "You get your money back in four months! What other product pays for itself like that?"

Edward's hoping his investment in the 1–2 Flush will more than pay for itself. So far, he's flushed $75,000 into it. "I sold my house so I could do this. Now I live in my shop, but my patent attorney is living well. I believe you've got to sacrifice. It's like they say in the weight room: No pain, no gain."

The product does require users to make a small change in their lifestyle. As Edward says, "'Oh my goodness, I have to hold the handle down for two or three seconds. What a hardship!' But when you get your water bill, you'll be happy. And when your well doesn't go dry and your water's in the well and not the septic system—that's peace of mind."

Sure, the 1–2 Flush has been a financial drain, but Edward hopes it pays off before he tanks.

Tilt-A-Roll™

"It's Your Turn"

STAT BAR

PATENT: US #5588615 and 5690302

PRODUCT PRICE: $24.95

STATE: Texas

INVENTOR'S AGE: 43

INVENTOR'S PROFESSION: engineer

MONEY SPENT: $15,000

MONEY MADE: $10,000

WEB ADDRESS: curtisbattsonline.com

"... most males are 'unders' and most females are 'overs.'"

It's a question that married couples and roommates have battled over for years. Should toilet paper be hung so it comes over the top of the roll or under and down the wall?

Curtis Batts may have found the solution, and who knows how many marriages he has saved. Curtis came up with a toilet paper dispenser that rotates. Just twist the dispenser and you can change the paper from over the top to along the wall. Each toilet paper user can now decide which way he or she wants to get their paper.

An engineer by trade, 43-year-old Curtis looks at the world through a problem solver's eyes. He set those eyes in the direction of the bathroom after hearing the results of a radio survey describing how couples argue over the direction of the toilet roll. Chuckling to himself, the survey reminded Curtis that his parents argued over this same "tissue issue." In fact, when he was married, he and his wife battled over this paper problem, too.

By the way, Curtis says he and his former wife are typical users; he's an "under" and his wife an "over." According to his survey of 2,500 people, most males are "unders" and most females are "overs." And here's an interesting fact: some people are so sure their way is right that 25 percent will change the roll to their preferred direction when visiting friends' and family's houses. Would that be you? I hope not.

Anyway, back to our toilet paper tale. Curtis figured there must be a solution to this problem. One afternoon ten years ago, he rolled up his sleeves and came up with the idea in just fifteen minutes. It was easy for this engineer. A little swivel is all it took. The Tilt-A-Roll swivels for those who are "overs" and tilts again for those who are "unders." And it can even stop in-between, leaving the toilet roll sideways for those folks who just can't decide.

It took Curtis three tries to get the prototype right. Then he shared it with his friends and family. After they stopped laughing, they saw the real need for his invention. After all, "over" or "under" is a sticky subject, right up there with the "up" or "down" status of the toilet seat. Curtis says that some people specify the toilet roll direction in their wedding vows. Those are folks who really should get the Tilt-A-Roll as a wedding gift.

Curtis has put his bidirectional toilet paper dispenser on sale on his website and others. He handles the marketing himself—contacting homebuilders, attending trade shows, and has been a guest on twenty different radio shows, including overseas radio.

His biggest marketing coup was his appearance on *The Tonight Show with Jay Leno* in 1999. His participation in a trade show contest got him on *The Tonight Show*. He went to the 1999 INPEX convention (the world's largest invention show), entered the Tilt-A-Roll in a contest, and won third place for its appeal and simplicity. *The Tonight Show* was putting together a segment about new inventors, so he and the first- and second-place winners were invited onto the show. While Curtis demonstrated the Tilt-A-Roll on the show, Jay Leno shared that he is not only an "over," but will even change the paper at others' houses, saying if people don't have it right, they obviously don't know which way it's supposed to go.

Toilet paper dispenser . . . with a twist

After his appearance on *The Tonight Show*, Tilt-A-Roll sales were up.

Curtis has found that promoting his invention is a tough job, especially since he works full-time helping people repair their credit and obtain affordable healthcare as well as providing legal services. But the $15,000 that he, family members, and friends invested in the Tilt-A-Roll should soon pay off.

Curtis is confident that with a little more marketing, Tilt-A-Roll sales will be over the top, and Curtis will be on a roll.

Flat-D™

Silent but Deadly

STAT BAR

PATENT: US #6313371

PRODUCT PRICE: $12.95 to $19.95 reusable

STATE: Hawaii and Iowa

INVENTOR'S AGE: Brian is 47

INVENTOR'S PROFESSION: postal worker

MONEY SPENT: $41,000 in the beginning

MONEY MADE: steady growth and profits every year

WEB ADDRESS: flat-d.com

"I finally found a material used for British chemical warfare suits . . . "

The National Guard should be proud of Brian Conant. Not only has he tried to protect us all from gas attacks, but he told me his story and included his wife's gas attack. This guy has got guts.

It all started while Brian was in service with the National Guard in Hawaii. Wearing military protective clothing, a charcoal suit, during a simulated chemical attack, "I released gas and noticed that I couldn't smell any odor from it. Nor could anyone else." It wasn't until his wife, Myra, had a similar episode that a connection was made. "The lightbulb went off after she said, 'Too bad there's not something you can wear so you don't have to worry about the odor.' I kept thinking, 'How could I come up with a way to bring that idea into reality?'"

Brian started experimenting with the material used in the special charcoal suit he wore the day of the simulated chemical attack, and decided it wasn't consumer-friendly. Any perspiration would leave a black charcoal residue. But he wasn't discouraged. "I finally found a material used for British chemical warfare suits—lightweight, breathable, washable, extremely thin ($\frac{1}{16}$ of an inch), flexible, and residue-free." At that point, this inventor declared war on flatulence.

Now, most people either deny the existence of smelly pops or blame the dog. But Brian was flat-out serious about flattening out the effect of flatulence. He designed The Flatulence Deodorizer (Flat-D for short), a three-ply activated charcoal cloth pad that's worn inside underwear, sort of like a panty liner. It absorbs intestinal gas odor before it escapes into the air where others can smell it.

Brian got some help near his home in Miliani, Hawaii, and sent out product samples and a video to big manufacturers of hygiene products. They all passed on the idea, declaring there wasn't a need for Flat-D, or any product like it. In Brian's mind, that's pure denial. "If you're thirty years old and have never released gas, you would have burst by now." According to his research, the average person expels gas fourteen times every day. The daily total ranges from as little as one cup to as much as one half-gallon. And I personally know people who release even more than that.

Brian wasn't fighting the flatulence war alone for long. Frank Morosky, who lives in Iowa, came up with a similar idea at about the same time. When he did his research, he learned that Brian had beat him to the punch in the patent office, but Frank didn't disappear. Brian relates, "Instead, Frank contacted me and said he had a marketing background and lots of ideas for selling the product I'd developed."

Brian and Frank joined forces and together are working to make Flat-D as essential to hygiene as underarm deodorant. As their website states: "If you think you're alone with your dilemma, realize that over 20 percent of the United States population (60 million) suffer from one or more medical disorders that cause excessive flatulence."

The Internet has been a strong weapon in their battle against embarrassment. Says Brian, "People who have diseases that cause flatulence are reluctant to walk into retail stores, but on the Internet they're not shy about searching for solutions. I'd say we've sold a total 40,000 units in various sizes and packaging to people in forty-five countries and in all fifty states." Their line has expanded the Flat-D line to include products for people with colostomies, a special line just for women, and even a chair cover that masks the sound and odors of a person who does more than sit.

Logistically speaking, Brian and Frank have found ways to divide and conquer. Brian oversees manufacturing and bulk mailings of the product to Frank in Iowa. There, Frank takes care of fulfillment and customer service. They share the load when it comes to planning, marketing, and publicity.

They get constant confirmation that Flat-D is fulfilling a serious need. "People see what a profound difference it makes in their lives. They say 'I can go on a plane trip, and go dancing again. It's really improved the quality of my life. It's given me freedom.' Hearing that is the gratifying part about being an inventor."

Brian, Myra, Frank, and his wife, Sharon, are owners of Flat-D Innovations, Inc. Frank is the company's only full-time employee. Brian continues to work for the postal service. "We've been putting our kids through private school and our 7-year-old daughter is in grade school, so we still need to make money.

"We're showing profits and growth every year, so we know we're going in the right direction." Brian and Flat-D haven't won the war, yet. Noticeable gas attacks are still very prevalent. But if Flat-D is not the ultimate solution, at least it's a product that doesn't stink.

Making gas seem like a breath of fresh air

Toothbrush Holder™

Hold On, Toothbrushes

STAT BAR

PATENT: US #6776296

PRODUCT PRICE: projects $8–$10

STATE: Georgia

INVENTOR'S AGE: 32

INVENTOR'S PROFESSION: salesman of commercial fire sprinklers

MONEY SPENT: $13,000

MONEY MADE: nothing yet

WEB ADDRESS: "would be cool, but not yet"

"I would classify it as a vision from a higher power."

Some dream of fame, others dream of fortune. Joe Herren dreams of dental hygiene. You see, one night Joe woke up out of a dead sleep at 3:00 a.m. with the idea for an invention—the Toothbrush Holder. He describes his experience this way: "Within seconds of my eyelids opening, I crawled out of bed, grabbed the closest paper, which happened to be our electric bill, and scribbled feverishly with a nearby pen.

"After about thirty minutes of jotting notes and drawing a crude sketch, I felt sure I had described in detail what my mind saw while sleeping. This was not a typical dream. And with my southern upbringing, I would classify it as a vision from a higher power."

It's sort of comforting to know that Joe's higher power is so concerned with toothbrush storage.

Anyway, the next morning, Joe explained his idea to his wife, April. She was so excited about it that she encouraged him to pursue a patent. He spent the next year looking for a trustworthy local patent attorney and researching the market.

You may be wondering why a higher power felt a new, improved toothbrush holder was necessary. Well, it's because the old wall-mounted toothbrush holders with a small center hole for a cup don't work with today's thick, rounded toothbrush handles that are all about grip. Joe knew his Toothbrush Holder was a nice-looking update and since they didn't find anything like his Toothbrush Holder, they went for a patent. Three and a half years and $10,000 later, Joe received his patent. That was his signal to get to work on a prototype.

It took a few weekends and several attempts for Joe to make his first prototype out of solid oak. Then he found out about a machinist in nearby Loganville, Georgia, who made a plastic version in four days. "Having a professionally made prototype in my possession, I started looking into manufacturing costs and potential profitability. The cost of manufacturing a unit ranged from 42 cents to 50 cents a unit. Local retailers were selling comparative products for an average of $7. This idea could potentially make money." Dreaming of cleaning up, Joe wasn't prepared for the next bit of news his research uncovered.

His big-money dreams were brushed off when he learned the cost of the plastic injection molds were $5,000 to $10,000 each, depending on whether they were made of aluminum or steel. The real holdup is that cleaning up with the Toothbrush Holder would require ten molds before a manufacturer would even consider production. Including materials, that put the start-up costs at $150,000. That number wiped the smile right off Joe's face. "At the time, my wife was seven months pregnant with our first child. Even with equity in our home and land, stocks, and excellent credit, I was nervous about the risk." And if you think Joe was nervous, that's nothing compared to his in-laws. They thought the idea was nuts. They're not big risk takers. And now that Joe and April have a baby son, Gavin, the in-laws really can't get behind this expense.

For now, the Toothbrush Holder is on hold. "I'd like to see them on the shelf, like any inventor. It'd be nice to get an investor to go in with me, so I'm not taking the full brunt of it. Exposing the idea is my next step. So here I am, telling my story to you, with the hopes my story catches the eye of a serious investor."

If Joe finds that serious investor he'll have a reason to smile and he'll be able to put toothbrushes in their place.

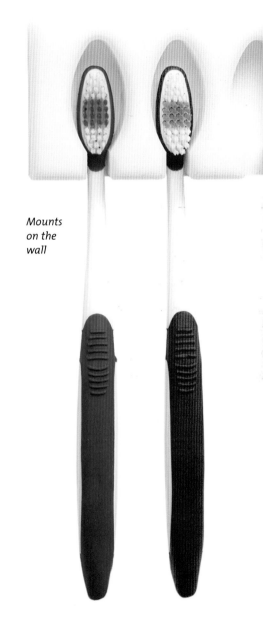

Mounts on the wall

So far, an unfulfilled dream

Bird Diapers™

Poop Patrol

STAT BAR

PATENT: US #5934226

PRODUCT PRICE: $19.99 plus various options

STATE: Virginia

INVENTORS' AGES: 46 and 47

INVENTORS' PROFESSIONS: both had Navy careers

MONEY SPENT: $10,000

MONEY MADE: $1.9 million in gross sales

WEB ADDRESS: birddiaper.com

Mark put his foot down—and, yes, stepped in some poop.

This one is literally for the birds. But if you've got a parrot or cockatoo, like inventors Mark and Lorraine Moore do, you'll see why this gadget will pamper you and your bird.

Here's the poop—whoops, I mean scoop—on how this invention took flight. The Moores had four wonderful birds named Sunny, Sara, Sam, and Rainbow. With four birds come lots and lots of poop—they go every fifteen to twenty minutes. So let's see, that's four times an hour and ninety-six times a day. Well, you do the math.

Lorraine liked to keep the birds out of the cage to enjoy their company, but the house ended up looking like a bird coop, or should I say bird poop. While Lorraine loved her fine-feathered friends and was willing to look the other way, husband Mark put his foot down—and, yes, stepped in some poop.

Then, an idea dropped from heaven. At a prayer meeting in San Diego, Lorraine and her friend Cely, a seamstress, had an idea to stop the poop. What if you put diapers on birds? As they talked, suddenly the light-bulb shined brighter and brighter. The next thing they knew, they were in Cely's garage, stitching diapers—for birds. Mark loved the idea, too, and his competitive spirit—a product of being one of eight children—fueled his desire for bragging rights to being the first of his siblings to get a patent for poop. His parents would be so proud.

Patent or not, the birds had to wear the diaper without, shall we say, getting

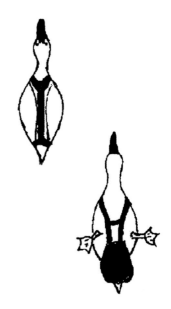

Stylish flight suits

their feathers ruffled. The first prototypes had metal snaps—an engineering no-no. Lorraine recalls the weight of the diaper causing the birds to walk like "drunken sailors." Their friends got a good laugh out of the spectacle, but the birds were not high on the idea. So it was back to the drawing board. Finally, the magic of Velcro produced diapers that did not inhibit preening . . . or pooping. The diapers were dubbed Flight Suits, inspired by Mark's career in the navy.

But our bird tale doesn't end there. No pet stores or catalogs wanted to market the diapers. And the bird enthusiasts lashed out at the product for being inhumane. Birds are not people, they argued. But that didn't stop the Moores on their crusade to make the world a cleaner place. On a wing and a prayer, they launched an Internet site and now sell over 1,200 Flight Suits a month from their now poopless home in Virginia.

HeadBlade
Toenail Trimmer
Healthy Hands
ActionCaps
Strapper
Pak Kap
Workout Wake Up
Smittens

You At Your Best

We all care about our appearance. We buy gel for our hair, moisturizer for our skin, and stylish clothes to fill our closets.

We want to look our best, and we are willing to spend our money to do it. From a sharp gadget that helps bald men shave their heads to a tech tool that helps older people trim their toenails, garage inventors have come up with ways to maximize our good looks.

In this section you'll meet inventors who have found a way to improve the baseball cap, hide bra straps, and even enhance our nipples. Did you even know your nipples might need enhancing? Two inventors certainly had

romance on their minds. One created winter mittens designed for a couple to share—as in two hands in one mitten. Another inventor came up with a sexy bow, so a woman could literally gift-wrap herself for the man she loves. Ah, yes, romance.

Whether it's skincare or ready-to-wear, these gadgets make us look or feel better. These inventors are counting on some basic human emotions here. There are strong forces at work behind our vanity and our desire to attract love. Inventors hope those forces are powerful enough to open wallets. Can you feel the love?

HeadBlade®

Heads Up

STAT BAR

PATENT: information not provided

PRODUCT PRICE: $15.00

STATE: California

INVENTOR'S AGE: 38

INVENTOR'S PROFESSION: graphic artist

MONEY SPENT: $200,000

MONEY MADE: millions

WEB ADDRESS: headblade.com

"Exposure and publicity are more important than retail space. If your product is sitting on the shelf and people don't know what it is, it doesn't matter—they'll walk right by it."

When Todd Greene moved to Seattle from Philadelphia, he decided the time was right. He'd been going bald, and moving to a place where no one knew him was the perfect opportunity to take control of his thinning hair and shave it off. That was 1993 and Todd was 26 years old.

In the early '90s, being bald wasn't as cool as it is today. There was no Vin Diesel, and Bruce Willis still had his hair. In those days, it seemed like only skinheads shaved above the ears. Still, once Todd started shaving his head, he found he really enjoyed the look. He also soon found that being light-skinned with dark hair meant he had to shave at least every other day. And it was time-consuming, taking twenty to thirty minutes to get the deed done.

After shaving his head for a few years and putting his art degree to work as an online manager for ESPN and as an artist for Sierra Online, he was rubbing his head with his palm one day when an idea hit him. Todd's shaving chore would go so much easier if he could hold the blade in his hand. Like holding a pencil near the lead, holding a razor near the blade would give the user more control.

After all, shaving your head is tougher than shaving your face.

Todd moved to Santa Monica, California, with his savings and his sharp idea. He made fifteen prototypes before one made the cut. Then he worked with a company to solidify his design and make a professional prototype of the HeadBlade. "I didn't create a blade," he says, "I created a comfortable handle with ergonomics. The HeadBlade's back pad creates suspension, so you don't have to worry about angle—you just drive it." People he knew who shaved their heads test-drove the HeadBlade and called it a smooth ride.

Three investors came on board: Todd's father and two of Todd's friends. Together, the four put up $200,000 and spent the first $50,000 on molds and production. Todd put everything he had into it, including his art background. He designed the artwork, website, and promotional materials. "I knew if I could get online commerce and national press, I'd be able to make a go of it. You don't need retail presence to sell something. Exposure and publicity are more important than retail space. If your product is sitting on the shelf and people don't know what it is, it doesn't matter—they'll walk right by it."

Todd took care of the online commerce piece and *Time* magazine jump-started the national press coverage. *Time*'s article "Ten Best Designs of 2000" legitimized the HeadBlade. Since then, it's been featured in dozens of magazines, on TV, and in several books. In fact, the design is so good that The Museum of Modern Art, MoMA, has included it in its collection. People can see or buy this museum-quality piece online or from

any of 12,000 stores that it's sold in, including Rite Aid and Sav-On Drugs.

Todd worked out of his apartment for the first two years and didn't take a salary. "For the first three years, I lost money. I started production in the U.S. But once I got the volume up, I realized that people don't care if a product is made in the U.S. or in China. They just want something that works and is cheap." Todd now owns 91.5 percent of the company—a company whose sales have doubled every year. It's now a multi-million dollar business.

Todd sees his product as part of a new lifestyle, coining the term "headcare," an offshoot of hair care. "We're branding the idea of active lifestyles merged with headcare. Many athletes shave their heads to gain a performance and psychological advantage. The brand represents active people who say, 'I proclaim who I am—I don't take life as it comes to me. I choose life on my terms.' These people are independent, but part of something bigger."

The second year Todd was in business, he learned that people losing their hair because of cancer treatments were responding to his

Motorcycle/jet ski design meets the razor

branding, too. Instead of feeling like a freak or being embarrassed, having a bald head could be a fashion statement. "If you know people who are going to lose their hair, get them a HeadBlade. It gives them a slant that being bald is cool—and they're proclaiming it. These products communicate that a lot of people must love having a shaved head." So now Todd's company partners with cancer organizations. His message? "Be concerned about things that matter, and not about just some hair." He's breaking the stereotype and stating that being bald can be empowering. "If you're bald, people don't know if you're sick or you're stylish."

Todd took a sharp idea and carved out his niche by heading toward his goals, facing his competitors, and going for the top.

Toenail Trimmer™

The Agony of De-Feet

It's a sad fact: time is no one's friend. As we get older, simple acts like bending over to cut our toenails become difficult. Roy Martin's arthritis made this grooming chore a pain in the neck, and everywhere else.

"It's simple—not at all complicated or imaginative."

He says, "I started experimenting with a better way to do my toenails because it was getting harder for me to reach down there, as it is for most older people sporting a bit of a gut. I had to put my foot up on the desk—there's a lot of strain. And my arthritic hands made it difficult to operate a traditional nail clipper. I kept putting it off as long as possible." This 74-year-old retiree stopped putting it off. Instead, he put his best foot forward with a step in the right direction.

Roy looked for a convenient tool to mount at the end of a baton. He wanted something motorized and tried every tool he could think of, including a chopper, grinder, and scissors. Then he tried some of Dremel's sanders, abrasive tools, and even its small saw blade. None of them worked. Roy was about to give up when he tried a dado blade. A dado blade rotates and is used in making fine furniture. It was a cut above anything he had tried. This rotating blade was mounted on a baton so that its teeth oscillated across the surface. It worked. Fine cuts yielded smooth toenails.

Roy's Toenail Trimmer is a baton about two feet long with a Dremel-type machine at the handle that drives the cutting blade. He says, "It's simple—not at all complicated or imaginative. But the design was sufficiently unique to get a patent."

This retired Air Force and National Guardsman from Glenburn, Maine, knew his invention had nailed it.

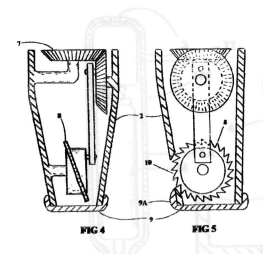

FIG 4   FIG 5

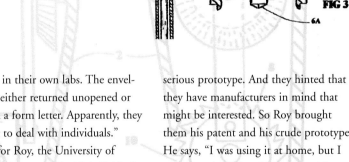

"I imagine with proper sales, it will be useful worldwide. And because it's so convenient, almost everybody will want one. There's no doubt that it works."

Almost everybody who sees the Toenail Trimmer likes it, and he hopes he'll "get the thing made." "I never told my wife, Elizabeth, what I was doing until I was preparing to go to the Yankee Invention Expo. But she, the family, and our friends all now seem to like the idea and want one." Roy has discovered that nurses are excited about using it to help them trim patients' nails, especially in nursing homes.

He tried to contact the major companies that he thought would be interested in his Toenail Trimmer, but none of them have showed any interest. "I've FedEx-ed many CEOs and executives by name, but companies don't seem interested in discussing products not originated in their own labs. The envelopes were either returned unopened or resulted in a form letter. Apparently, they don't want to deal with individuals."

Lucky for Roy, the University of Maine does want to deal with individuals. The people at the Advanced Manufacturing Center take pride in helping local Maine residents develop products like this. They took the Toenail Trimmer on as a project, having students develop the serious prototype. And they hinted that they have manufacturers in mind that might be interested. So Roy brought them his patent and his crude prototype. He says, "I was using it at home, but I gave it to the university to play with as a guide to making the final product. So now I'm back to suffering."

Let's hope Roy doesn't have to suffer too long before his Toenail Trimmer makes the cut.

Healthy Hands

This Inventor Hopes to Clean Up

Here's a dirty little secret. Kids hate to wash their hands. And when you do get them in front of some soap and water, they don't wash their hands long enough to kill the germs.

This is something Rick Ruskin has been aware of for a long time.

Rick has three children who are ten, eight, and four years old. Rick's wife, Julie, has made hand-washing a priority in their household.

Over the years, parents have all told their children to say their A-B-C's or sing Happy Birthday twice while washing their hands to guarantee the hand-washing takes long enough to do the job. But those parenting tools don't often work. Have you ever heard how fast a child can say their A-B-C's when they're really motivated to finish it quickly?

Rick Ruskin has worked in the toy industry for years—mostly on the marketing side. But the hand-washing conundrum pushed Rick into the design and invention world of kids' products.

The idea for hand-wash timing wasn't all that new, but Rick wanted to improve upon the existing concept. He and his design partner, Mark Hartelius, started to focus on how the timer could be integrated into the natural action of dispensing soap. Rick's toy sense kicked in and he knew that the magic of fun characters, music, and lights would help to motivate kids. The "ah-ha" moment came when they designed a timer that could be attached to almost any existing soap bottle. This way consumers could stick with the soap brands they trusted and attach this timer directly to their current dispenser.

As with most inventions, Rick's musical timers hit a few sour notes along the way. Rick knew that there were plenty of toy companies that look to outside inventors for new product ideas. He liked the idea of being paid a royalty on sales for something that someone else was making, so he found a company willing to do just that. The company

paid for the tooling and the production of musical computer chips. It seemed like a perfect marriage, until the company decided to change focus and dropped the product line that included Rick's product.

Now Rick had the product tooled and he had 50,000 chips that played "If You're Happy and You Know It." (Rick's version is "If you're healthy and you know it, wash your hands.") It was the beginning of 2008 and the process was stalled.

About three or four months later, Rick had a chance meeting with a distributor of Disney promotional products.

"I showed him the timers to see if he would be interested in selling them. His response was that if I incorporated the Disney licensed characters and Disney music, then this was a sure thing."

For those of you thinking that a Disney license is easy to come by, think again. It took a few months, but ultimately Healthy Hands and Disney had a finalized deal. For girls, the Disney version would be The Little Mermaid, which plays "Under The Sea"; and for boys, it would be Lightning McQueen from the movie *Cars*, playing "Real Gone," the opening song from the movie.

In June 2009, two things happened: Rick lost his day job, and he decided to manufacture the minimum-order quantity of 24,000 musical timers to jump-start the business and capitalize on the concerns over swine flu. Healthy Hands was now his major focus.

Rick launched his web site during the last week of August 2009, and that same week he was featured on NBC's *Today Show*.

About two months later, Healthy Hands was featured on the nationally syndicated talk show *The Doctors*.

"It was really exciting to watch the orders online the day the segment aired. Early on, we were excited to see four or five orders come in online on any given day. By the end of the week after *The Doctors* aired, we had 250+ orders."

Rick had a significant success selling Healthy Hands online. In three months, he sold about 2,500 units. Those impressive sales numbers helped get Healthy Hands into a few retail stores, including Ralph's Supermarket chain on the West Coast (part of Kroger's) and into Babies R Us. Rick is currently exploring getting Healthy Hands into other chain stores.

If it all goes as Rick hopes it does, then lots of kids will have clean hands and Rick will clean up.

ActionCaps™

Keep the Cap, Change the Brim

This inventor's lightbulb started to blink while he was in Malaysia. Casey Robinson and his friend were walking in downtown Kuala Lumpur when they looked through a store window at a U.S. music video that was playing. A band member on the video wore his hat sideways and his friend asked why people do that. Casey answered, "Everybody wants to stand out and be different. They should make one hat with different brims that can come off so they can switch them."

So in addition to bringing home souvenirs, this traveler also brought home an idea.

The development of ActionCaps was a gradual evolution. The first prototype had nuts and bolts on the sides. It was a good cap for Frankenstein, but not for the general public. Several prototypes later, Casey perfected the design with bungee cords. He gave one to a friend who wore it to the opening game of the Giants' season. Because his friend really liked it, Casey decided to patent it in 1992. His patent for the first baseball cap with an aftermarket was issued in 1993. Now he sells a variety of brims and crowns made from various fabrics in different colors, with an option to have a message embroidered on the brims—perfect for rallying a team with the brim straight up displaying a message like Home Run, Rally Time, or Touchdown.

Casey has been inventing part-time for thirteen years. During that time, he's earned a degree in electrical mechanical engineering. He currently has a job with a manufacturer that makes regular baseball caps. The owner liked the way Casey was pursuing his own cap business and hired him to activate his customer service.

STAT BAR

PATENT: US #5253364

PRODUCT PRICE: $16–$20

STATE: Colorado

INVENTOR'S AGE: 43

INVENTOR'S PROFESSION: electrical mechanical engineer, customer service rep for cap company

MONEY SPENT: $200,000

MONEY MADE: $15,000

WEB ADDRESS: rdzyn.com

While he hasn't made a home run, he's made a few base hits.

Before getting married seven years ago, Casey had put everything he made into this cap venture. But being a smart guy, he cut his spending back after marrying Rhonda. Overall, he'd spent a total of $200,000, maybe more. Good friends who didn't want to be repaid have helped him along the way.

Casey concentrates his cold calling sales on corporate sponsors for baseball games. His pitch? If you give away a product, at least give something the fans will bring back to the game. So far, this 43-year-old hasn't gotten his caps into the right hands (or on the right heads). While he hasn't yet made a home run, he's made a few base hits. He'll go three months without hearing anything, then get an order for 140 pieces.

ActionCaps give you brim flexibility.

He doesn't attend trade shows as much as he used to, but he does continue to do mailings and contact advertising and promotional companies. He's looking for someone else to run the bases for him. Not sure why companies haven't stepped up to the plate, he chalks it up to a matter of timing. "I have to be upbeat. I break down every few years, but I just pick myself up and get going again."

His biggest problem in approaching baseball teams is that most of them have contracts with Nike or another corporation, so a player can't even wear a different cap. "I can't get past the gatekeeper. A year and a half ago, I sent every baseball team in the country a sample of a rally cap. Only twenty contacted me, but said they couldn't do anything about it because they already had a contract with Nike or Reebok. When the teams have a contract, that's what every player has to wear. This pervades other sports, too. The official hat for the 2006 Superbowl was Reebok. And take Tiger Woods. Before he started his own company, he got paid to wear Nike. I don't have any money to give him to wear mine."

The effects of "big business marketing" runs deep. Casey saw its underbelly first-hand at the biggest apparel trade show, called the Magic Show. It cost him $6,000 to get there, and he's still paying for it two-and-a-half years later. He garnered some attention walking around with the ActionCap on. Everybody was looking at it. A representative from Majestic West, a corporate apparel company, said, "I like your hat, but I can't do anything about it because New Era (a cap manufacturer) pays me to not make baseball caps. They pay me very well. If New Era doesn't jump on it, there's nothing I can do."

While Casey's ActionCaps aren't getting much action yet, he's staying at bat. There's no capping this player's enthusiasm.

...how about a piece that would reinvent your own bra...?

Strapper™

Bra-vo!

How does a girls' night out, a fashion faux pas, and a cut-up credit card equal invention inspiration? Read on.

One fateful night about a decade ago, now-34-year-old Michelle Ostaseski went out for a night on the town at a local Long Island, New York, bar with her girl-friend. Into the bar walked a group of girls who, Michelle says, were "wearing racer-back tank tops with not a care in the world about showing their bra straps." That's right, they had exposed bra straps.

Michelle was "stunned" by the women's blatant fashion faux pas, which was eliciting stares and dirty looks from other ladies at the bar. Of course, the men at the bar didn't seem to mind as much.

"It turned out that, by the end of the night, we'd found these girls to be quite nice." Michelle wished she could've said something to them, "but how do you bring up the subject? It's like when people have something in their teeth. I didn't want to embarrass them." Pass the pretzels and, by the way, your bra strap is showing.

Riding the Long Island Rail Road train home from her job in Manhattan's Garment District, Michelle couldn't get the visible bra straps out of her head. "So I started drawing on my take-home paperwork. I thought about something that would bring in the straps. Should I reinvent the bra? Or how about a piece that would reinvent your own bra, but not for good?" Michelle needed to find a solution to this bra bugaboo, so she turned to the ultimate provider of tank tops and bras—her credit cards.

Michelle experimented by cutting up her old credit cards, and creating a device that pulls a woman's bra straps together in the back. Doing that hides the bra straps from view when a woman wears a sleeveless tank top. She called her invention The Strapper.

That's where this story stopped. "I sat on the idea for eight years. I didn't know who to tell. Plus the cost of a patent at that time seemed like a million dollars." And Michelle was strapped for cash.

Eight years later, she met her future husband, Jimmy. "He pushed me," she says. "Jimmy knew I was frustrated with my current job at a department store and he encouraged me to get the patent I always talked about. Unfortunately I still didn't know how to go about getting a patent, so I went to Invention Submission Corporation, the worst possible choice ever!"

Issuing a warning to other inventors, Michelle explains that she was charged $8,000 and told they would find a manufacturer for The Strapper. "Two years passed, and nothing."

A frustrated Michelle started doing her own legwork. Her phone calls to manufacturers whom ISC had contacted on her behalf proved shocking. "Most of them were no longer in business, and some weren't even plastics manufacturers."

Michelle was devastated but also determined. After a thorough search, she found the right manufacturer, one who listened. "I've been with this company ever since."

When her first prototype came back, Michelle test-ran it on her mom, Juana. The results were a letdown.

"My mom was uncomfortable with the first Strapper. She said the 'large' was too small. I was so mad at her for crushing me, but then I thought, 'If she's uncomfortable with it, an average-sized woman, then I have to change it.' When the new prototype arrived, my mom loved it! And thanks to her, so do so many other women."

After launching her website, www.thestrapper.com, and booking her first store order at a boutique called Girlfriends in East Norwich, New York, Michelle's business was off the ground. Later the next year, she attended her first trade show. "The product did great! I started booking trade shows around the country."

The Strapper is now in about 200 boutiques, Michelle reports, and she is currently contacting cheerleading organizations, since "safety pins are now banned from competitive use and straps being seen is a reason for disqualification." Rah, rah, go Strapper!

Michelle is quick to point out the Strapper would never have happened without the help of her family. Her husband's push resulted in applying for a patent, and her mom's frank feedback improved the design. Her sister, Sandra, and brother-in-law, Scott, also helped with the packaging. There's no question, The Strapper is a family affair.

All indicators show that The Strapper will be a success. Michelle Ostaseski has given women the support they need and, at the same time, created a hidden asset.

Bye-bye bra straps

STO DESIGNS™
The Strapper
The Bra Accessory

Pak Kap™

Using Your Head

Only apple pie is more American than a baseball cap. Baseball caps have been around more than 140 years. To try to change the baseball cap would be like trying to change the American flag. Who would attempt to alter this American icon? Arnie Pedersen.

STAT BAR

PATENT: #D503261-S

PRODUCT PRICE: $22

STATE: Texas

INVENTOR'S AGE: 57

INVENTOR'S PROFESSION: textiles

MONEY SPENT: $80,000

MONEY MADE: "still in the red"

WEB ADDRESS: pakkap.com

"Untold is unsold. So we continue to keep pushing and trying new avenues."

About three years ago, Arnie came up with his heady idea while jogging in the park near his Houston home. Arnie is not a marathon runner. He only jogs occasionally. But he uses his jog time wisely. While most of us jog mindlessly or maybe think about telling off our bosses, Arnie was thinking about—his keys.

Here's the problem. Arnie would drive to the park, lock his car, and then jog. But what about his car keys? There wasn't a good place to put them. Arnie's shorts were lightweight and made jogging with the keys uncomfortable. Holding the keys in his sweaty hands was not a good solution. Tying the keys inside his shorts was also uncomfortable. Where could he store his car keys? Arnie scratched his head, and the idea for Pak Kap was born.

Pak Kap is a baseball cap with a handy storage pocket designed into the back of the cap. Because Arnie mainly jogged with a cap on, it seemed a good location for the keys. Once he had a working prototype, he realized that at the gym, he could use his cap to store his membership ID and locker key.

One day Arnie found he had nowhere to store his sunglasses while jogging. He then added the sunglass loops to hold his glasses in place.

Arnie had no problem cutting and sewing caps. He was in the business of making privacy curtains and drapes, so he had a commercial sewing room. He also had an embroidery business with access to manufacturers of caps. With these two resources, he was in the perfect position to reinvent the cap.

Arnie partnered with John Edmondson, a friend from Pittsburgh he had known for over twenty-five years. Both men had a textiles background.

They made a series of prototypes until they settled on one. Then they passed out sample caps to friends, family members, and a few strangers. The feedback they received was totally positive, except for one person—Arnie's 24-year-old son, Erik, who didn't like it. Erik thought the cap didn't look cool. Arnie appreciated his son's opinion, but it didn't stop him. He also knew that if his cap were sitting on a superstar's head or it had a Nike Swoosh on its side, it would be cool.

Arnie got his patent and registered the name Pak Kap. He found sources overseas to produce it. Arnie and John didn't want this cap to blend in with all the other caps on store shelves, so they decided it should be sold in a box. They also added cheap sunglasses to bring attention to its glasses loop.

Arnie presented his Pak Kap to two retail chains and they both bought it. Trial orders went into sixty stores. It looked like Arnie's baseball cap was ready to score a homerun.

Unfortunately, it ended up being a pop fly to center field. The Pak Kaps sold slowly and both retail stores needed a faster turnover.

Arnie has continued to sell his Pak Kaps at small specialty stores.

He did get a crack at QVC, selling about $2,500 worth of merchandise per minute. Unfortunately, QVC requires a minimum of $5,000 per minute.

Since then, Arnie and John have been working the corporate market, making Pak Kaps for companies such as Miller Brewing, New Process Steel, Beihl Shipping and others. They have invested about $80,000 into their cap fund. In the first year, they sold about 6,000 hats.

Arnie says they are still in the red, but not discouraged. "Sales look brighter. We are getting some recognition now and repeat sales—still not enough to recover our expenditures, but we aren't discouraged. Untold is unsold. So we continue to keep pushing and trying new avenues."

Arnie sees many untapped markets to pursue. He knows that wade fishermen will like his Pak Kap because it will keep their fishing license high and dry. Golfers will like it for storing their ball markers and extra tees. Beach-going vacationers can use it for their room keys, cash, and credit cards. And Arnie points out another untapped market—nudists. After all, "they have no pockets."

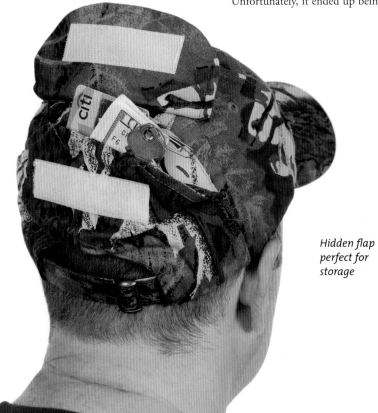

Hidden flap perfect for storage

Workout Wake Up

An Alarming Exercise Program

STAT BAR

PATENT/TRADEMARK: pending
My Wake Up Call® has a registered
trademark. My Workout Wake Up
Call™, My Wellness Wake Up Call™,
Motivational Alarm Clock™, My
Wake Up Calls™, My Wellness Wake
Up Call™ are in their final stages of
becoming registered marks.

PRODUCT PRICE: $19.99 for month of
8–10 minute messages on MP3; files
for iPods/iPhones/MP3 players, or CDs;
$44.99 for My Wake Up Call® CD and
iPod Alarm Clocks (includes one
month of messages); $59.99 for My
Wake Up Call® Cube Alarm Clock (with
stored messages inside the clock)

STATE: California

MONEY SPENT: $40,000

MONEY MADE: "Approximately
$10,000 at this point"

WEB ADDRESS:
http://www.mywakeupcalls.net

"This is your wake up call."

The road to invention can start anywhere. For Robin Palmer, it started on Broadway.

Robin performed in Broadway musicals for 15 years. After leaving the stage in 1994, Robin pursued her passion for fitness and became a personal trainer.

Eventually Robin moved to California and left the world of fitness for insurance marketing. The money was good, but compared to Broadway and personal training, insurance marketing is not a very physical job.

Robin fell into the typical suburban rut. She sits for a living and after dinner she'd watch TV; then she would go off to bed.

Robin missed her more physical life, and she wished she could afford a personal trainer to motivate her to wake up and work out.

One morning in 2004, Robin's alarm went off just as it did every morning, but this time it triggered an idea: "Imagine if your alarm clock woke you up with an inspiring message that helped you start your day more productively."

Robin created My Workout Wake Up Call™ Motivational Alarm Clock, featuring daily motivational messages designed to help listeners to wake up and start their workout.

The CD alarm clock

The iPod alarm clock

Turning this idea into a reality was not as easy as falling out of bed.

Robin has spent a lot of time, money, and energy on getting trademarks, copyrights, and a patent. Her patent lawyer first charged her about $20,000. A year later, he wanted an additional $5,000. Robin exercised her right to complete the patent process on her own.

Next, Robin needed to record her motivational messages. A friend knew a sound engineer who worked for a radio station and volunteered at the neighborhood high school. So, after meeting with the engineer, Robin ended up recording her messages in the Westlake Village High School band room utility closet.

To get the music for her messages, Robin had an inspiring idea. She turned to her dad, who happens to be Robert Boudreau, the conductor of The American Wind Symphony Orchestra. Her dad donated the rights to many music pieces that Robin uses in her messages and, since her dad often works with Pete Seeger, she also got access to several songs from Seeger.

Next, Robin built a web site to sell her motivational messages. She hoped visitors would purchase the messages and then load them into their existing alarm clocks.

That didn't happen. Instead, consumers asked Robin to sell the clocks, too.

"It was a real challenge finding compatible clocks—and finding them at an affordable price. I started in Broadway. Now I'm in world of electronics—it's been a steep learning curve."

In May 2009, one of Robin's friends was in New York City and just happened to ride the elevator with an editor at *Shape Magazine*. Being a good friend, she used the elevator ride to pitch Robin's product to the magazine editor.

The pitch worked. My Workout Wake Up Call™ was *Shape Magazine's* featured product of the month in the August, 2009 issue.

The article caused Robin to get dozens of calls and E-mails. And on the web site, she sold about 120 units. Robin said it could have been more, but the shopping cart feature on the site had problems and many sales were lost.

Right now, Robin has the task of putting together the packaging for each outgoing product sold. Robin says it takes about 30–40 minutes to package each alarm clock.

"My dining room table is my shipping and branding department."

Robin is looking to find ways to streamline the process. She also wants to improve the clock and make it her own design.

Robin is very tired, but is still moving forward, proving her motivational alarm clock must be working.

The MP3/cube alarm clock

Smittens™

Love Is Cold

"Why can't we hold hands inside the mittens?"

Brrrr. It's cold outside! Your hands are frozen, your body chilled, and your nose is leaking God-only-knows what. But yes, despite that arctic chill, love is in the air. And what could be more romantic than a walk in the bitter cold with your honey?

A romantic walk requires hand-holding, but wait a minute—you've got gloves on. Gloves holding gloves—that's no fun. As we all know, true romance requires skin-to-skin contact. My winter wonderland romantic walk has been ruined! Don't fret, my love muffin. You need some Smittens. No, that's not a typo, it's Smittens, as in mittens built for two who are actually "smitten." Why not? There are two-seated bicycles and love seats, so why not Smittens? Can't you just feel the love?

Inventor Wendy Feller conjured up this lovely idea after many years in the clothing design business. She started out in New York City designing sweaters for some of the top-name department stores. Unfortunately, there was no love fest there and she packed her bags and moved. If Wendy had moved to Florida, the story would have ended there. But Wendy moved to Seattle. She obviously likes cold environments. Her first foray into invention was a personalized blanket. You would send her a picture of, say, your loved one, and she would convert the image to black-and-white pixels. Then she'd stitch the image onto a blanket. It was a labor of love that ultimately was too labor-intensive.

One cold and blustery day, she was strolling with her husband, Allan. The cold didn't stop them from holding hands. But trapped inside mittens, it felt very awkward. "Why can't we hold hands inside the mittens?" Wendy asked her husband. It was at that magic moment that Smittens was born (actually reborn, because Wendy had had this idea years ago but never followed through).

First, Wendy tried a prototype with one opening for both hands. That was mitten mayhem. Then she came up with the idea of trying two openings. The second opening made all the difference. Holding hands in the winter would never be the same. Using polar fleece material, she sent some homemade samples to magazines, which featured the product. Uncommon Goods, a specialty catalog, warmed up to the idea of Smittens and placed a large order.

Wendy, now smitten with success, is in her third season of Smittens. Love triumphs again.

Together forever?

Headwarmer Pillowcase
1st Class Sleeper
The Slanket
Boo-Boo Blankies
CozyHold
Vidstone
Sun-Cap Floating Sun Shield
KosherLamp
Kling Tut
SkyRest
CandleWand
SnuggleTite Slumber Bag

At Rest

2:23 . . . click . . . 2:24 . . . lying in bed, watching the clock as the minutes just crawl by—it's torture. Where's Mr. Sandman? We need our sleep, and more importantly, we all want to sleep. But in a world where we spend our lives running around, slumber is often a luxury.

For those tired of staying awake, the multi-billion dollar pharmaceutical industry has come up with all sorts of pills and potions to help us get some shuteye. But rest assured, garage inventors have stayed awake nights coming up with their own sleep-friendly solutions. Hello, beddy-bye.

One inventor from Maine has a pillowcase designed to keep the top of your head warm on a cold winter's night. Two other inventors have soared to new heights to help us catch a few Z's while flying on commercial jets. When the baby can't sleep, neither can the parents, so another inventor has come up with a blanket a baby can't kick off. There's even a Kosher Lamp that lets Orthodox Jews rest knowing they are not breaking any religious laws when they turn off their bedside light. And for that final eternal sleep, one inventor has added video technology to gravestones.

Normally we say, "You snooze, you lose." But with these inventors, if you snooze, they win. They're all working so you can sleep easy. Pleasant dreams . . . click . . . 2:25 . . .

Headwarmer® Pillowcase

No Head Case Here

STAT BAR

PATENT: information not provided

PRODUCT PRICE: $14.95–$34.95

STATE: Maine

INVENTOR'S AGE: 58

INVENTOR'S PROFESSION:
general contractor

MONEY SPENT: $130,000

MONEY MADE: "more than $30,000,
but still in the red"

WEB ADDRESS: headwarmer.com

*She knocked on doors, talked to
people in parking lots, and
approached strangers at local bars.*

Is it possible to improve upon the pillowcase? Julie Brown thinks so. In fact, Julie is so sure, that she's has invested over $130,000 of her savings into her new pillowcase design. Julie's product is a cozy flannel pillowcase with a built-in head blanket, designed to keep its user's head and neck warm during cold nights.

Julie Brown, an inventor at 58 years young, spent her last thirty years in construction. Starting as a secretary for a Los Angeles builder, she got pretty good at reading plans and blueprints, and worked her way up the company, ultimately taking the general contractor license exam—the only female in a room of 900 men, most much younger, taking the test. Julie passed, and worked as a general contractor before starting her own company specializing in remodeling restaurants and churches and doing tenant improvement work in San Diego. On retirement, Julie decided to live full-time in her motor home.

Julie's invention came to her on Thanksgiving weekend in 1998. She was visiting friends in New Hampshire when it started to snow. Her friend, Sara, con-

cerned about Julie sleeping in the motor home, wanted her to use the guest room. This story would have ended there, except Julie's aging standard poodle couldn't make it up the stairs in Sara's home. Sara still fretted that her friend would be cold, but Julie responded that only her head would be cold. The remark caused Sara to go on about her own husband, Roy, who spent each night diving under the blanket to warm the top of his head, then pulling the blanket back because he couldn't breathe under the covers. This went on all night long, every cold night, and all that blanket activity kept Sara awake.

That night in the motor home, while trying to keep her own head warm, Julie kept thinking about Sara, Roy, and the blanket problem. Then it came came to her in a vision. According to Julie, she could see the Headwarmer pillowcase with "flashing marquee lights all around the image." She jumped out of bed and sketched the image. Julie's life would never be the same again.

Back in Rhode Island, Julie began working on her idea, making ten prototypes and sending most of them to Roy to test. Then she took a class for inventors and entrepreneurs at The Center for Design & Business at the Rhode Island School of Design (RISD). She filed for a patent and hired a graphic artist to help her create a logo.

Much of Julie's effort revolved around finding the perfect material for the pillowcases. It had to be soft to the touch, but with enough body to wrap around the head properly, not droop into the user's nose, mouth, or eyes. The material also couldn't just lay on the face. If you moved it, the wrapping had to stay where you put it. Julie ended up finding a 100 percent cotton flannel with a double nap (fuzzy on both sides). It's actually a rare find; the special flannel isn't sold in fabric stores since it's only used to line silverware drawers and for massage table covers.

The next step was to really test the Headwarmer pillowcase. At this point, only Julie and Roy had tried out the prototypes. Now Julie needed a larger test market. To hear her tell it, Julie went out and approached strangers and convinced

them to try her new pillowcase. She knocked on doors, talked to people in parking lots, and approached strangers at local bars. I'm not sure if asking a stranger in a bar to try your pillowcase is the smartest move, but it worked for Julie. She signed on 100 participating testers. Each study participant got a free Headwarmer pillowcase with the understanding that they had to fill out Julie's survey. The results were amazing: 98 percent really liked her product. That was the encouragement Julie needed.

Today, Julie runs her pillowcase business from her motor home, primarily traveling between Arizona and Maine. She submitted her invention to the National Mail Order Association (NMOA), nmoa.org, for their Made in America Hot Product Contest, a national search for the most unique and interesting

Sweet dreams

products. Julie's pillowcase was chosen as the winner for the State of Maine.

In the first three years, Julie has sold 1,250 pillowcases (with only one return) at about $25 each. Considering that she has invested $130,000 of her own money, her sales have been less than ideal. But Julie's not giving up. Since learning that mostly men (80 percent) buy her pillowcases, she's dropping the Chill Frill name in favor of something more manly and producing a line of headwarming pillowcases with a camouflage print for hunters and fishermen.

Ever since she had that late-night vision of her pillowcase with flashing marquee lights, Julie's been sure she has a hot idea to stay warm at night. I guess only time will tell if Julie's dream comes true.

1st Class Sleeper®

Fasten Your Seat Belts

"When I started this, I wasn't an inventor and wasn't out to make a product. I was out to get some sleep."

TravelLady magazine's review of the 1st Class Sleeper said, "I could pay for a coach ticket and feel like I was flying in first class." The reviewer was talking about an inflatable travel pillow that lets you do the impossible: sleep well—in coach.

Bob Duncan, an airline pilot for Alaska Airlines, commuted to work—a two-hour drive to Seattle and a 3.5-hour flight to Anchorage (as a passenger, not the pilot), followed by his work shift. Then came the tough part, the commute home after being on duty for twelve hours.

Bob had to find a way to get some rest during his homeward flight. Being a big guy, at 6'1" and 230 pounds, it was no easy task. He tried stretching across three seats and surrounding himself with pillows, but he still couldn't sleep. One day, he stuffed eleven pillows in his lumbar and upper back area and used the seat belt to hold himself in place. Two hours later, he woke up. Bob had found the rest he was looking for, and realized he might just have landed a first-class idea as well.

Bob took the idea and flew with it. His first prototype was alarmingly colorful. It consisted of a purple beach ball topped by a ball the size of a grapefruit, wrapped in a yellow life vest and held in place by gray duct tape.

"When I started this, I wasn't an inventor and wasn't out to make a product. I was out to get some sleep. People on flights looked at me with my funky-looking, multicolored prototype. But they'd noticed how well I'd been sleeping during the flight and wanted to try it— and then they wanted to buy it. I wasn't about to sell my prototype—it was the only one I had. But their interest really surprised me."

It took some nudging from above to get this 44-year-old pilot into the seat cushion business. When Bob and his wife, Mary Jane, went on a mission trip to deliver medical supplies, they left the travel pillow behind. After the long flight home, they were both feeling tired and achy. That's when Bob felt that God spoke to him, saying, "I gave you this idea and you haven't shared it with anyone. If you won't run with it, I'll give it to someone else." This pilot had been on auto long enough; it was time to take orders from the control tower in the sky.

At a family meeting, the Duncans agreed to go for it. It meant sacrificing a college fund, but that money has since come back to them. They've sold 30,000 pillows so far, having spent $130,000 to lift this idea off the ground (including a costly advertising venture).

Early on they had decided they would advertise where people feel less than first class—that would be coach. They picked *SkyMall,* the onboard catalog magazine, as a good way to wake people up to the Sleeper. And it was—except for the tragedies of September 11 which happened during the six-month advertising run and cost them $44,000. Now the Duncans have ads running in various airline magazines, and soon in *American Way,* American Airlines' in-flight magazine.

First class comfort at coach prices

Looking for more venues to sell the Sleeper, they had approached Magellan's Travel Supply Catalog twice and were turned down. After a trade show nightmare in Chicago (the 1st Class Sleeper lost to a can of ham!), Bob asked the judge if she'd sat in the Sleeper. She said, "No, but I walked by it."

A fast learner, Bob next went to the the Super Bowl of luggage trade shows put on by the Travel Goods Association. This time, he stipulated on his application that he didn't want to be considered for an award unless a judge actually sat in the Sleeper. He watched as the judges argued over who would sit in it. Finally

one sat down. He immediately closed his eyes and said, "I'm not getting up." The judges then fought over who got to sit next, and the 1st Class Sleeper ended up being awarded the prestigious "Momentum Award." One of the judges happened to be from Magellan's so, shortly after, the travel supplies megastore decided to give Bob's invention a test flight. Today, the Sleeper—with purchases exceeding 100 a week—is Magellan's number one selling travel pillow.

Sales have been climbing 40 percent every year for the last three years without a nosedive in sight. Bob can sit back in comfort and watch his sales soar.

The Slanket®

Sleeve Me Alone

"She wanted me to have both hands free to read, instead of watching TV all the time."

No library, no studying for Gary Clegg. His first semester in college was a breeze. Every night he would stay up late watching TV and still pull down straight A's.

Gary was experiencing the perfect freshman year. That is, until he hit December. You see, Gary was going to school in Maine and December was freezing. Even his dorm room was bitter cold, but that didn't stop him from massive TV viewing. Gary would cocoon himself in a thick blanket and try to channel surf through the blanket. Unfortunately, his roommate's remote wouldn't send a signal through the blanket. Gary had a problem. So one especially cold night while watching Conan O'Brien, completely covered with a blanket up to his chin, Gary opened the blanket so he could stick out the remote. The cold air rushed past the remote right to Gary's skin. Brrrrrrr. Gary had a big problem. The chill was enough to get him moving. Gary grabbed his roommate's buck knife and cut a slit in the blanket, just big enough to slip the remote through.

Not attempting a cover-up, Gary confessed his couch potato behavior to his mother. When he returned home during winter break, Gary asked his mom to sew a sleeve into his blanket. He thought that would solve his remote problem. To his surprise, Gary's mom made two sleeves. Gary explains, "She wanted me to have both hands free to read, instead of watching TV all the time." That's a good mom. His two-sleeve comforter became Gary's favorite blanket and he used it through all four years of college.

"I brought it everywhere. At the time, I traveled to Colorado and Utah to compete in snowboarding. Friends tried to convince me to develop the blanket-with-sleeves idea and market it. But I was still trying to figure out what I wanted to do with my life. I wasn't confident enough to start a business. I had no idea where to start."

See, the Slanket lets you do more than control the TV remote.

Immediately after college, Gary moved to Brazil to teach English for a year. It was a life-changing experience that pumped him full of self-confidence. "I was thrown into it. I had to be a professional, command a room, and teach. In college, I could procrastinate. In the real world, I was forced to prepare. That's when I became motivated."

After his year in Brazil, Gary moved in with his brother in Boulder, Colorado, to write a novel. "For ten months, I lived on the couch and wrote." Gary's favorite blanket went with him. He called his invention the Slanket because it's a blanket with sleeves. I guess it could

have been called a Bleeves, but Gary went with Slanket.

During this period, Gary and his mother perfected the Slanket design and started selling Slankets to family and friends. "Mom always believed in me. She'd do whatever it took. She'd send me a check for $150, or for whatever number had been sold. Our orders were growing, just by word-of-mouth."

Gary next moved to New York City in hopes of pursuing an acting career. Gary didn't land any significant acting work, but he did catch New York City's entrepreneurial spirit. "While I don't like the rat race culture I see in New York

City, where people are stressed out and money is the motivator, being here gave me the last push to look at the financial piece. I figured, 'Okay, let's do this. Everyone else can do this. I can, too.'"

He called his older brother, Jeff, who encouraged him to pursue the Slanket. They figured out the start-up costs. "I had some money saved up, but not enough. Jeff offered to help me out. Now, with my brother's money invested in it, I had another motivation to forge ahead."

It began like a marathon. Gary conducted research like he never did in college: textiles, trade shows, manufacturers, and legalities. He set up his

company and recruited help to create a hip website. His brother, a software designer, set up the site and a friend from high school designed it. "When I flew to North Carolina for a textile show, that's when I knew I was serious. I was going on a business trip!"

He bought fleece from a company in China and found a factory in Maine. "I couldn't have my mom make them all." He wanted to launch his winter product in October, but didn't get it all together until late January. They missed the holiday season. "I spent months biting my nails. I just wanted to sell enough to pay my brother back. I was prepared to sell door-to-door if I had to.

"When we launched the website, my brother and his friend Jay posted Slanket references on blogs and websites all over the Web. Within three days, the Slanket was mentioned on 200 blogs. It infiltrated the Internet." Orders went from zero to seventy, in one day. Then nytimes.com, *Wired* magazine, and a magazine in Ukraine contacted Gary. "The first week, I sold one to Alaska, Honolulu, Helsinki, England, and France. I couldn't believe it!"

He started with an inventory of 600 and sold out in five weeks. "Every weekend, I'd drive from New York City to my parents' house in Maine to package Slankets. I recruited two friends and bought them pizza and beer. We turned packing nights into parties."

Gary and Jeff easily made their money back and a profit, to boot. QVC contacted Gary three weeks after they started. Now the National Football League wants to license it, and department stores have contacted him.

"When I first started, I knew I wanted the company to be an extension of myself, because the product is an extension of my lifestyle. I like it when companies have a personal feel. I want to deliver a really good product to people—that makes me happy. And I want other people to like the Slanket as much as I do.

"Customer service is really important to me. The money is secondary. When I packaged the Slanket for those five weeks, I wrote a personal thank-you note for each one I sold. I got an incredible response. That's the stuff I'm really into. It makes it fun. I'd rather be like that and make decent money than have no personal connection and make millions.

"This whole thing is like a big school project, but there's money involved. That's the only difference. The money. I haven't seen it. It's only numbers in the bank, but it lets me buy more supplies."

Gary started as a slacker and ultimately made the Slanket. He took his love of TV and his hatred of the cold and came up with an idea that's keeping him warm at night. No one is going to throw a wet-blanket on this guy's dreams.

Boo-Boo Blankies®

Boo-Boo Blankie, Bye-Bye Tears

"After the Boo-Boo Blankies were on Regis and Kelly, all the local news stations interviewed me, and the newspaper put in a feature about me."

Sometimes a strong-willed toddler can be the spark of inspiration. After Krissy McCoy's son, Jake, had surgery on his eye, he needed to keep a cold pack on it for twenty-four hours. Not liking the cold, this toddler made it clear that no cold pack was going to touch his face.

Krissy says, "Jake's never liked cold rags on his forehead. He doesn't even like me to take his temperature. I thought if I wrap the cold pack in something soft, maybe he'd like it better. He loved fleece blankets, so I stitched a pocket in a piece of fleece, and put a cold pack in the pocket." The blanket worked like a charm. Jake healed like a champ and the Boo-Boo Blankie was born.

Krissy's friends thought it was a smart idea and maybe a good way to make money while staying at home to raise kids. Let's face it; kids are accidents waiting to happen. Just as a skinned knee or elbow heals, it gets bruised again. The Boo-Boo Blankie is Krissy's answer to any minor boo-boo, and that includes earaches and tummy aches. It's a security blanket that's small enough to take anywhere and doesn't even drag on the floor.

"I wasn't a seamstress of any sort when I first started. Before I made it exactly how I wanted it, I sewed at least twenty." The trick was figuring out how to get the fabric to stay straight in the satin ribbon while running it through the sewing machine, since fleece has some "give" and wants to pull. Krissy also had to decide where to place the pocket, determining what worked and looked best.

Once she had the design the way she liked it, Krissy started the process of getting a patent. That's when she got a boo-boo. She learned that her product is not the type of invention that's patentable. Even though each one is monogrammed, appliquéd, and infused with love (her secret ingredient), it's not

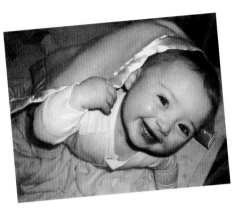

unique enough—there are other blankets with pockets on the market.

Krissy was down, but not out. Instead of getting a patent, Krissy trademarked the name "Boo-Boo Blankies" and warmed the hearts (as well as the boo-boos) of kids around the country.

To get Boo-Boo Blankies rolling, Krissy spent more time than money. Searching for a way to work at home, she used a tax refund to buy a computer and taught herself how to build a website. It took her two weeks to get it up and running. The timing was perfect. A week later, Kelly Ripa used a Boo-Boo Blankie on the morning TV show *Live with Regis and Kelly* to soothe Regis when he was sick.

In a surprise smart marketing move, Krissy had sent a Boo-Boo Blankie to one of Kelly's children on her birthday, hoping Kelly might pass the word to her friends. To Krissy's surprise, Kelly took the Blankie on the show, then shared Krissy's website with the world. "With a forty-five-second snippet, I got the majority of my orders that one day!" Then, before Christmas, Kelly brought it back on the show as one of her top ten favorite gift ideas. That brought in more orders, and sent Krissy shopping for an upgrade to her sewing machine. "After the Boo-Boo Blankies were on *Regis and Kelly,* all the local news stations interviewed me, and the newspaper put in a feature about me." Gotta love Regis and Kelly. Krissy sure does.

Another goal of Krissy's has been to start a charity called "Close to My Heart" for children with serious illnesses. She donates one dollar from the sale of each blanket to a child in need, including Hurricane Katrina victims and those struggling with cancer.

Even when they're not sick, Krissy's kids still use their Boo-Boo Blankies. "Jake will be five this summer, and he likes to hide his toy dinosaurs or plastic men in the pocket. My little girl, Macy, uses hers to snuggle up with."

And Boo-Boo Blankies aren't just for kids. Older people with arthritis appreciate the soft covering for their therapeutic hot/cold packs. People have even ordered Boo-Boo Blankies for their pets. Someone ordered one for a graduation gift, and one lady ordered a throw-sized blanket with five pockets in it; she said she wanted to be warm all over. Just like boo-boos, Boo-Boo Blankie users come in all sizes.

So far, Boo-Boo Blankies haven't healed Krissy's financial boo-boos. She's had to go to work as a dental assistant to keep her home finances from unraveling. Krissy hopes the Boo-Boo Blankie business will one day allow her to hire a few stay-at-home mothers to help, and go back to being a stay-at-home mom herself.

In a world of bumps and scratches, Krissy McCoy is making it "all better" one Boo-Boo at a time.

A sort of cuddly way to make boo-boos all gone

CozyHold™

Hold Me

STAT BAR

PATENT: pending

PRODUCT PRICE: CozyHold, $29.99;
CozyHold Pro, $39.99

STATE: Connecticut

INVENTORS' AGES:
Randy, 26; Marcia, in 50s

INVENTORS' PROFESSIONS:
business owners

MONEY SPENT: $100,000

MONEY MADE: sold a few thousand
since March 2003

WEB ADDRESSES:
ramarindustries.com & cozyhold.com

"We came up with an idea of a holder that makes it easy to hold cold or hot compresses."

Wisdom teeth and a package of frozen peas inspired mother-daughter team Marcia and Randy Nozik with this business idea.

I'll let 26-year-old Randy Nozik describe what happened: "About five years ago, I had all my wisdom teeth pulled at once—all *five* of them. . . . I was in a lot of pain and kept pressing packages of frozen peas against my jaw. My hands got numb from holding the frozen package, plus the peas started to smell as they thawed out." Not a pretty situation.

"Not long after that crisis passed, my mom and I were brainstorming ideas at the kitchen table about a business I could start after college. We came up with the idea of a holder that makes it easy to hold cold or hot compresses." Good-bye, frozen peas. Hello CozyHold and CozyHold Pro-gel pack holders, made of neoprene, that use Velcro to strap onto the area needing thermal attention.

Randy is a tennis player, and always has to ice her knee for about twenty-five minutes after playing. She also knew that many people with a host of ailments could also benefit from holding a gel pack tightly against their sore spots. "Our ideal target market is large. It includes people over fifty who manage chronic pain like arthritis. It includes athletes in any sport at any age. And it's great for families with young children. If a little boy falls off his bike, his mom or dad can soothe him immediately by wrapping CozyHold on his bruise."

Back at the kitchen table, Randy and Marcia made prototypes out of muslin that would hold two gel packs in two different sizes, and then sent them off to a manufacturer. It took three tries to get it right. Once the design was the way they wanted it, they began manufacturing with a company in China. The gel packs, which come with the CozyHold neoprene strips, are purchased in quantity in the U.S. and the packaging is printed here, too. Mother and daughter fulfill orders out of their home office in West Hartford, Connecticut. In the meantime, a patent lawyer has submitted their thirty-page patent application full of graphs and diagrams for approval. "Right now, CozyHold is patent pending. We've already got the champagne on ice to celebrate when it comes through," says Randy.

As most inventors have told me, the toughest part of being an inventor is marketing. The Noziks say they would like to advertise in *AARP* magazine, but the cost would break their budget. CozyHold has been highlighted in *Bottomline* magazine as a new product, and that generated sales. So has advertising there. "We're striving to get publicity in women's health and fitness magazines, and catalogs in the U.S. and Canada. It takes luck, uniqueness, and the right connections to get into the big catalogs like Harriet Carter and Sharper Image."

Randy and Marcia did work with a company that promised to produce an infomercial. But they found out it was a scam. Says Marcia, "It's important to research these companies really well. We learned that they keep changing their names and playing the same scam on new businesses."

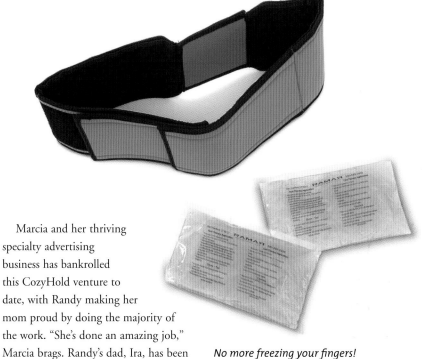

No more freezing your fingers!

Marcia and her thriving specialty advertising business has bankrolled this CozyHold venture to date, with Randy making her mom proud by doing the majority of the work. "She's done an amazing job," Marcia brags. Randy's dad, Ira, has been highly supportive, too. A professional photographer, he did the product photos that appear on the cozyhold.com website for their company RAMAR Industries, LLC. The "RA" is for Randy and the "MAR" is for Marcia. Clever, huh?

After three years of pressing to sign up with the right distributors and selling a few thousand CozyHolds off the website, Randy is ready for a hot streak. Promising is a businessman from Korea who is close to arranging distribution in Korea and Japan. "We've been changing the packaging to suit that market. We're looking for a similar kind of distributor here."

Marcia, however, especially savors one small victory. "I was in a local drugstore and saw a woman wearing a CozyHold on her head, as if nursing a toothache. I chased her down in the parking lot and asked where she got it. The university dental clinic had recommended it, and this woman had nothing but praise for her CozyHold, saying she had no swelling at all after surgery because of it."

So far, the response to CozyHold has been both warm and cool. But this mother-daughter team is sure that, with a little more pressure, the CozyHold phenomenon will soon take hold.

Vidstone®

Final Cut

STAT BAR

PATENT: pending

PRODUCT PRICE: $1,999.00 retail
(sold through funeral homes)

STATE: Florida, moving to Colorado

INVENTOR'S AGE: 33

INVENTOR'S PROFESSION:
telecommunications executive

MONEY SPENT: "close to 7 figures"

MONEY MADE: not yet available

WEB ADDRESS: vidstone.com

"I want the name Vidstone to become the generic name for multimedia memorials"

When we visit the graves of loved ones, what do we do? Well, we try to remember the dearly departed, picture their face, try to recall their voice, attempt to communicate on some spiritual level, and stare at the gravestone. Gravestones never really reflect the person that's gone. They're just markers. Where's the personality, the wisdom, the smile, the sense of humor?

What if a gravestone could really capture who you are, or were? That's Vidstone, a tombstone with a built-in video, so visitors can see and hear you in living color after you're gone. No, you can't take it with you but, thanks to Sergio Aguirre, you can leave behind on your gravestone a video starring yourself.

In March 2004, Sergio attended his father-in-law's funeral. He noticed how the somber mourners livened up as they watched a slide show tribute about the deceased. They giggled seeing a photo of him with a martini in hand, a pink boa wrapped around him, and a top hat on this head. At the interment, Sergio scanned the gravestones that stood as lasting tributes to the departed. That's when the idea struck him dead on.

Why shouldn't people be able to view a 7- to 10-minute high-quality tribute at a loved one's gravesite? The vision for Vidstone—video stone—came to life.

Sergio researched companies serving the funeral industry with multimedia products and found "funeralOne." Sergio's idea was to make it easy for this technology and consulting company's video presentations, customized by local funeral directors, to be permanently placed outdoors in a cemetery. The funeral corporation had the software for creating video memories. Sergio only had to come up with hardware durable enough to withstand temperatures as high as 140°F (60°C) and as low as -40°F (-40°C), and withstand sleet, rain, snow, hail, winds, and everything else.

A few months after deciding to make lasting graveside videos a reality, Sergio created a prototype Vidstone Serenity Panel. Its screen is covered by a solar panel, which can be flipped open by visitors. Once opened, the video starts. The device includes two standard headphone jacks to listen to the audio. The solar panel protects the screen from sun damage and charges a battery inside. Four hours of sunshine provides enough juice to play the video for up to ninety minutes. Its solar battery lasts about four years, making it relatively maintenance free throughout its lifetime.

Sergio's first Vidstone prototype won an award for the most innovative product at an International Cemetery and Funeral Association convention (ICFA) in Las Vegas. He took his prototype to the next level, then showed it off at the National Funeral Directors Association meeting in Chicago the following October, and

again at the 2006 March ICFA show. It killed. That's when he knew this could be something big. The ICFA convention is the largest international industry event for burial products and services. It represents 7,000 companies in the cemetery and funeral services profession. Winning the award and additional accolades told Sergio his customer base—funeral and cemetery directors—saw an everlasting value in his product.

"I want the name Vidstone to become the generic name for multimedia memorials, just as Kleenex is for tissues," Sergio says. There's at least one indication that could happen. In a television episode of *CSI New York* called "Fair Game," lead actor Gary Sinise is staring at a video on

a tombstone and says, "Ah, this is one of those new Vidstones." Dead-on product placement, huh?

Sergio has sourced high-quality suppliers to make the components of Vidstone—the first 1,000 assembled in April 2006, the same month he and his wife, Christina, relocated to the Denver area and the same month his daughter was born. "It's been a whirlwind since we started this a year-and-a-half ago."

Sergio has even bigger ideas for Vidstone technology. "We'll soon be able to retrofit Vidstone Serenity Panels onto existing gravestones. When the word gets out, the demand will be high."

And let's face it; for those who want a better way to remember loved ones at their place of rest, these Vidstones are heaven sent.

Afterlife TV

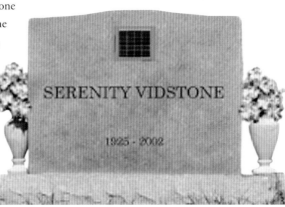

Sun-Cap™ Floating Sun Shield

Sun Block

"I wouldn't realize how sunburned I was getting until it showed up the next day."

Bruce Capwell, like most of us, has a love-hate relationship with the sun. He loves the sun—it hates him. Burns, wrinkles, and the risk of skin cancer all make the sun a dangerous lover.

"I'm one of those people who likes to stay out in the sun all day." Bruce is a painting contractor who loves to lounge in his Greene, Rhode Island, backyard pool. "I would fall asleep on a raft in the pool, and I wouldn't realize how sunburned I was getting until it showed up the next day. My wife, Diane, was always on me to stay out of the sun."

Bruce wanted it all. He wanted to lounge in the pool and he wanted to keep the sun away from his tender skin.

With both of those thoughts burned into his brain, Bruce headed off to

Home Depot. He bought various supplies, went into his workshop, and emerged as the inventor of the Sun-Cap, a "floating sun shield."

The Sun-Cap not only solved Bruce's sun exposure dilemma, it also generated lots of interest from friends and family who saw him lounging in it.

"Everyone wanted one," says Bruce, a 63-year-old father of two grown sons and grandfather of Megan, 7, and Jacob, 8. "Six months later, I filed for a patent, which was issued a year later. Then I got a trademark."

The original Sun-Cap really burned Bruce when it turned out to be too long for UPS to ship. Ouch, that hurts. "It wouldn't go through its automation system. So my wife had the idea to make it inflatable. I found an outfit in China to manufacture it, and then they suggested that I also get a design patent."

All these patents and trademarks turned out to translate into costly attorney fees. "When I think about how much I've spent, it makes me sick," says Bruce, who estimates he's invested between $70,000 and $80,000 in the Sun-Cap.

Despite the mounting bills, Bruce believes the Sun-Cap will generate big sales, "especially in the Sun Belt, where people can't even use their swimming pools during the day because the sun's so strong."

Bruce got a taste of how big the Sun-Cap could be when it was selected by The Skin Cancer Foundation to be featured in its journal, which goes out to 85,000 doctors, dermatologists, and libraries.

Bruce took that success and went on the road to pool-supply stores. He got the Sun-Cap stocked in stores in Rhode Island, Massachusetts, and Long Island, New York.

But a conflict arose. "The pool stores are closed in the winter; they're open from March to November, which is also my busy season as a painting contractor."

Since Bruce is unable to sacrifice the time to make the rounds to retailers, he concluded, "I need a partner. I'm looking for the right person with the right connections. If the right guy could see the product, I just know it could be huge. Unfortunately, I have to make a living."

Until the right salesperson sails along, Bruce is enjoying sunburn-free summers, while he keeps his dreams for his Sun-Cap afloat.

All the fun without the sun

KosherLamp™
Oy, Do I Have a Lamp for You!

To appreciate this invention, you need to understand something of the laws and customs of Orthodox Jews. I could go on about *melacha* and Exodus 35:2,3, but here's the bottom line. On the Sabbath (Friday sunset to Saturday sunset), observant Orthodox Jews are not permitted to do certain things, such as turn a light bulb on or off. They're not allowed to rip toilet paper either, but that's not relevant here.

STAT BAR

PATENT: N/A

PRODUCT PRICE: $35.95

STATE: Canada

INVENTOR'S AGE: 51

INVENTOR'S PROFESSION: Rabbi

MONEY SPENT: not disclosed

MONEY MADE: not disclosed

WEB ADDRESS: KosherImage.com

Keeping the above in mind, Rabbi Shmuel Veffer was asked by his wife, Chana, an avid reader, if there was some way to hook up a bedroom light so she could read there on the Sabbath. It's a problem Orthodox Jews have faced for thousands of years. Well, actually only since electric lightbulbs became popular.

Some Jews have gotten around this problem by using timers or just leaving the lights on. But Rabbi Shmuel wanted to come up with a better solution. The Rabbi says he's a trained problem-solver. "That's why I became a rabbi."

After looking through the Jewish legal sources, he came up with an idea. The lamp could stay on throughout the Sabbath, if it had a moveable shade so the user could turn the lighting on and off without turning the lightbulb on and off.

So Rabbi Shmuel and his 15-year-old son Shalom went to a Home Depot. They came home with Styrofoam, louvers, vents, electrical tape, lightbulbs, ballasts, wires . . . and who knows what else. Using the louvers, they developed a rudimentary, clunky, but working model, and tried it out the following Sabbath. Everyone liked it, but the consensus was that it needed to be smaller. The rabbi went back to the drawing board.

As if by holy intervention, one of Rabbi Shmuel's congregants was in the lighting business—a match made in heaven. After showing the working model to Moshe Lazer, the congregant

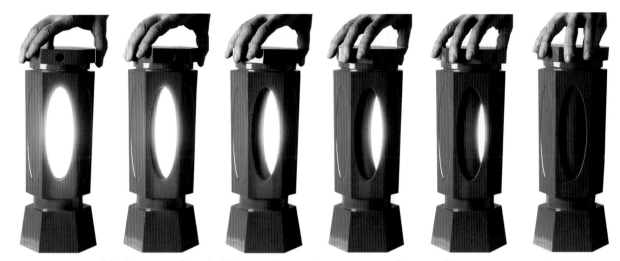

Turn the shade and you block the light—you don't turn off the bulb.

saw its potential and, after a handshake, KosherImage.com was born. No circumcision was needed with this birth.

One of the main concerns with early prototypes was heat. When you covered the bulb, heat would build up and the lamp became a fire hazard. The problem was solved by developing a zigzag tunnel inside the lamp. Heat could escape around the corners, but light could not. Next, they switched to a compact fluorescent bulb which produces less heat . . . and uses less electricity.

Moshe set off to China to build the prototype while Rabbi Shmuel continued to refine the lamp's design, getting input from potential consumers. Paperwork was soon filed by a patent attorney, a website was set up, and rabbis were consulted.

The initial model of KosherLamp was an instant success. Orthodox Jews saw the light, and liked it. "You can 'turn it off' in accordance with the laws of Shabbat. When you want to sleep, the 'fadeshade' technology allows heat to escape safely while virtually blocking the light—all without touching a light switch."

But is this lamp a kosher "loophole" or is it cheating? The rabbi explains it this way: "Resting on Shabbat is not about 'restrictions,' it's about bringing one to a higher recognition of God. KosherLamp is very much in the spirit of Shabbat. Every time I twist the KosherLamp shade, it's a way to think about God in a way that I wouldn't during the week when flicking the switch on and off."

The KosherLamp has become almost commonplace in observant Orthodox homes. It's even sold in more than a dozen countries on five continents. In Israel, the lamps are sold in eighty stores. In the U.S., it's sold in over 200 outlets, from Lubavitch Jewish Center in Anchorage, Alaska, to Holy Land Judaica in Boca Raton, Florida.

Rabbi Shmuel knows he's selling to a small niche market, but with 1.5 million Orthodox Jews worldwide, he sees the potential. He has also come up with a KosherClock. The alarm sounds for a period of time, then it turns itself off.

Rabbi Shmuel took a bright idea and brought it to light. And for Orthodox Jews, the Sabbath will never be the same.

Kling Tut™

It's Good to Be the Kling

STAT BAR

PATENT: US #6817032

PRODUCT PRICE: unknown

STATE: Illinois

INVENTOR'S AGE: 52

INVENTOR'S PROFESSION:
mechanical engineer

MONEY SPENT: $50,000

MONEY MADE: none

WEB ADDRESS: klingtut.com

He wanted the Kling Tut to look fashionable.

Scott Hollander knows he's fat. And those extra pounds can be tough to carry around. But when you have to fit those pounds into a tiny coach seat on a commercial jet, what was inconvenient becomes torture.

"I have to take everything out of my pockets and try to squeeze myself between two armrests." His only comfort came from folding his arms across his chest, but after twelve hours his arms felt like they would fall off. He was developing tendonitis in his arms and shoulders. If he finally did fall asleep on long flights, his arms would drop to his sides—or worse, onto a neighbor's lap. That disturbed his neighbor, and woke him up as well. Needless to say, Scott hated to fly.

On a thirteen-hour flight to Japan, Scott thought, "If I had a way to support my arms in a fixed, but releasable position along my upper body, I could relax the muscles used to hold my self-embrace." A way dawned on him; a vest and hand mitts that attach by using Velcro. The vest is called Kling Tut because the self-embrace pose recalls the

Egyptian King Tut. He modified "King" to "Kling" because of the Velcro action.

Scott inscribed the details of this idea in an inventor's notebook he's kept for years, and it entombed between its closed pages for three years. Then, excavated, he spent four years crafting prototypes. "I wanted to know the nitty-gritty, so I took sewing classes to learn what materials would make the lowest-priced product possible." He first learned to make a pillow, then graduated to a vest. An early prototypes looked like a referee uniform with Velcro strips running vertically along the front. He then found that, if he used velour for the vest, a mitt with Velcro would stick to it. The front of the vest was set.

The women in his sewing class unearthed indigo disco denim with a satiny luster for the vest back. Next, Scott designed fingerless mitts with

Velcro-lined palms. He wanted the Kling Tut to look fashionable, and then reveal its functionality when the wearer broke out the mitts. It had to be a garment people wear anywhere, and not look dorky.

With a prototype made, Scott planned on generating interest by wearing the Kling Tut while sitting near the boarding gates in the airport every weekend. But after the tragedies of 9/11, he couldn't get to the gates. Also, he'd planned to borrow money from his 401(k) plan, but the value went down and he no longer had enough to borrow against it. An invention marketing firm offered to charge $7,800 to do patent research and look for someone interested in licensing the product. Scott—like so many other inventors I've talked to—found out that many marketing firms are more interested in taking an inventor's money than making the inventor money. Scott ended up hiring a reputable patent attorney for $6,000—but that amount still grew to $20,000. Scott says he was able to talk the lawyer's bill down a few thousand dollars. If only he could patent that technique.

This 52-year-old mechanical engineer from Carol Stream, Illinois, has sunk $50,000 into the Kling Tut, including $7,700 to produce 325 prototypes. "I don't know how I've been doing it.

Functional, but not really fashionable

I took a major $7,000 loan against my 401(k) for the patent and raised $4,000 in seed money from two friends. The rest has come out of my pocket."

Scott is still working full time, purchasing farm equipment and using his purchasing moxie to negotiate the best deals for his invention. Scott's fiancée, Colleen, joined his efforts by designing the royal website for Kling Tut. While the rest of his friends and family thought his idea was creative, most weren't ready to "buy a brick" at $2,000 each. Scott developed a prospectus for those who are comfortable signing his nondisclosure agreement and is offering outsiders the opportunity to invest in the venture. So far, only two have enrolled.

Still, Scott envisions the Kling Tut being used in more than travel situations. Medical uses abound as a rescue or medical mobilization device. Maybe that one would be Sling Tut. Military personnel might find it useful for sleeping in bunks. The Air Force could call it Wing Tut. And then there's . . . oops, sorry. I got carried away.

Anyway, Scott wears it on airplanes whenever he travels. "No one on an airplane has asked me about it yet. As I keep my arms crossed with my blue neoprene mitts, I wonder if they think I'm an invalid."

The Kling Tut is not yet for sale. Scott would like to license it instead of running the business himself. He's entered it in contests for several TV shows including *Oprah* and *Made in the USA*—but so far no takers. His inventor's notebook is also full of other ideas, "But seeing how much I've spent on this, I need to have the Kling Tut hit the ground first."

Scott is sure his invention is like the real King Tut—buried treasure. Once unearthed and publicized, the lines will form.

SkyRest®

The Plane Truth about Sleeping in Coach

STAT BAR

PATENT: "yes, and trademarked"

PRODUCT PRICE: $29.95

STATE: California

INVENTOR'S AGE: 56

INVENTOR'S PROFESSION:
mechanical engineer

MONEY SPENT: not quite $50,000

MONEY MADE: recouped investment,
covers expenses, gets a small salary

WEB ADDRESS: skyrest.com

"The key word is persistence."

Globetrotting isn't easy. Long flights can be really tough. Forget about getting a decent sleep; it just isn't possible, right?

David Corbin knew the inside of cramped aircraft cabins all too well. During his former life as a design engineer for Xerox, David had to routinely fly back and forth to the Far East. Being a tall man, he also had to fold his six-foot-five frame into coach seats. It's a trick even Houdini couldn't pull off.

On one such long flight, David noticed some students who had propped their backpacks on the tray tables, and leaned forward to get comfortable, even snoozing a bit. As a trained engineer, he envisioned a product. "They needed something higher to put their heads on—something big and bulky yet soft." Of course, a pillow like that would be too big and cumbersome . . . unless it was inflatable. Since David couldn't sleep, he used that time to sketch his sleep-friendly idea. David designed the SkyRest—a pillow you can lean forward on comfortably.

To develop SkyRest, he decided to do as much of the work himself as he possibly could. That included doing his own patent research and application— "the key word is persist-ence"—plus handling the manufacturing, marketing, and even the distribution.

"In the beginning stages, I sent my drawings to several manufac-turers in China and Hong Kong, and kept corresponding by e-mail with the ones that responded. Eventually, one emerged as the best fit—a company that makes inflatable children's toys."

But would the traveling public spend money on a large blow-up pillow? David didn't rest easy until his first order for 1,000 SkyRests was placed. During the next four years, David sold about 10,000 units, mostly through one advertising resource—*SkyMall* magazine. "About a million air travelers a day browse through this magazine when they're flying, so that's helped make SkyRest a recognizable product." And let's face it; it's the perfect place for SkyRest to advertise. After all, you have a fleet of uncomfortable passengers who can't sleep flipping through the *SkyMall* magazine. They see the ad and it clicks. "The SkyRest ad gets into people's heads and they want it. A picture is worth a thousand words."

Hooking up with *SkyMall* has been a dream come true for David. He started out paying $11,000 a month for a quarter-page ad, but a high volume of sales helped him negotiate a partner deal. Now, he splits revenues from the sales that come through *SkyMall*.

David's side business of selling SkyRest turned into his "real job" when Xerox let go of 800 employees and forced David into early retirement. "I'm grateful that happened," David says. "I didn't think SkyRest could support me and my family, but I'm now covering my expenses and paying myself a modest salary."

This 56-year-old entrepreneur spends an average of three hours a day running his business out of his San Jose, California, home. Most of that time, he's fulfilling orders that come through SkyMall and his website. But for him, the money is secondary.

"I'm a musician. I love to play the guitar, classical style and a hackneyed version of flamenco." He plays with a singing group at a nearby church. "After working mega hours for thirty years, I feel like I'm just waking up to a new life and asking, 'What have I been doing all this time?' Today, I'm getting paid in satisfaction."

His satisfaction, ironically, also involves international travel . . . but now for fun. "My wife of twenty-five years, Meena, is a travel-holic. We went to India together twice last year." Meena was born in India and was working in Silicon Valley when she met Dave. Her family still lives in India, and she and Dave may purchase property in a suburb of Bombay called Puna, a fast-growing center of high tech.

Naturally David and Meena take their SkyRest pillows with them on every trip. Designed to inflate and deflate quickly, they are easy to pack into carry-on luggage, and the pillow can be adjusted for all sorts of situations. When Dave and Meena traveled by train, for example,

Inflatable air travel pillow

Airline Pillow

Pasenger sleeping with inflatable pillow

Food Service tray

Airline seats

they made SkyRest the size of a regular bed pillow and used it that way. It can also be used as a footrest while waiting in airports or when watching TV, or positioned as a table or a desktop pillow for "power naps."

SkyRest's future appears to be high-flying, especially if David closes a few deals with distributors in Korea, India, Norway, and even China and Europe. "That would keep the factory orders high and get me out of the daily fulfillment operation."

For a guy who just wanted to get rest while traveling, this invention may just let David Corbin rest easier—for the rest of his life.

CandleWand™

Burn, Baby, Burn

STAT BAR

PATENT: information not provided

PRODUCT PRICE: $24.95

STATE: Oregon

INVENTOR'S AGE: 44

INVENTOR'S PROFESSION:
energy industry consultant;
rep. for lighting manufacturer

MONEY SPENT: $220,000

MONEY MADE: $100,000

WEB ADDRESS: candlesave.com

"I look at it like it's an opportunity, not like it's my baby."

After going through a tough divorce and moving her two young children from Tucson, Arizona, to Portland, Oregon, Jillianne Pierce needed to de-stress. Working two jobs, she was burning the candle at both ends.

"I bought a two-foot-tall, hand-poured organic candle, wanting to meditate. The second time I tried to burn it, the wick was sunk down in the wax, so when I lit it, only a bit ignited. A teeny-tiny flame struggled in this huge candle. As the wax melted, it flooded the wick and the flame went out. Frustrated, I got out a sharp knife and attempted to cut off the top and start over. Instead, I cut my hand badly. I was mad. I'm stressed and trying to relax—and this happens."

Sound familiar? Others might have given up, but growing up poor, Jullianne was used to coming up with solutions to problems. As a kid, she and her brother entertained themselves thinking of new ways to do things. She wrapped her hand and heated a butter knife to trim the candle. It worked, but the knife cooled down too quickly. So Julianne found a soldering iron. It cut, but its 900–1,100 degrees was too hot. "The wax was smoking and melting everywhere. It was a mess." So, less hot, and shaped more "like a butter knife—not sharp, but blunt and flat" spelled the beginning of the CandleWand—a tool that safely trims uneven candles so they burn like new. When her friends kept asking to borrow it, she knew she had something hot.

Jillianne was familiar with electrical products, having imported UL products and energy-efficient lighting from Asia and India for her jobs as a rep for a lighting manufacturer and as a consultant to the energy industry for conservation programs. Still, issuing a letter of credit for someone else's product is a far cry from designing your own product.

Jillianne found a manufacturer in China who created UL-approved curling irons and would use the same components on her product. That made getting through UL easy, once they figured out how to categorize the CandleWand. She had to make only two revisions—to the temperature and the handle. After she got UL approval, Jillianne designed the packaging herself.

Everybody was supportive. Her family said, "Great! Go, make us rich." But no one could show her the ropes. Jillianne kept looking for someone who had done it all—a mentor. She learned that inventors have to be careful about people who say they can do it all. She found that people specialize in banking, packaging, etc., but if they say they can handle every detail, they're probably only adept at handling your money— and making it disappear.

When Jillianne needed $40,000 to buy a containerload of CandleWands, she was lucky enough to borrow the money from family and friends. So far, she's invested $220,000 in her burning solution. But she treats it as a hobby, since she's working full-time and raising two kids, McCord and Zak, now ages 15 and 14. She says, "I look at it as an opportunity, not like it's my baby."

Still, she takes it seriously. She put the CandleWand through independent, third-party product testing at the Innovation Institute. The Institute used the Preliminary Innovation Evaluation System (PIES-X) to test CandleWand's marketability. It got a very high score—

Rehabilitating old candles

42 out of 50. "They said it's worth pursuing, but with a caveat. It's not a new kind of coffeepot—it's an entirely new product, so I'd have to educate the public." Jillianne's word to the wise: Don't light a fire under your invention unless you have money to educate consumers about it.

"By the time I got the product and approvals, I had no money for marketing. People won't know to go look for it; they won't know it exists. Even if they see it on the shelf, they won't recognize what it is." She got interest from a few major retailers who carry products from large candle companies like Yankee Candle. They loved her product, but didn't want it on the shelf next to their candles in case people

might equate its presence with a low-quality, unevenly burning product. Yet no matter how high-quality the candle is, other factors like drafts in the room influence uneven burn. Jillianne recently started selling her product in the Solutions catalog. She's had more success with Solutions than online, but not even a flicker of retail success.

It was meditation and relaxation that brought Jillianne to the CandleWand, but her experience selling it has been a slow burn. Yet this mother of two isn't giving up. She knows that when consumers see the light and understand what her product can do, her sales will be hot.

SnuggleTite™ Slumber Bag

It's a Wrap

STAT BAR

PATENT: US #7013507
(Canadian patent pending)

PRODUCT PRICE: $59.95 U.S. or $69.95
Can. (includes shipping & handling)

PROVINCE: Ontario, Canada

INVENTOR'S AGE: "grandmother"
(age not given)

INVENTOR'S PROFESSION:
entrepreneur owner/operator
of day care center

MONEY SPENT: "a lot"

MONEY MADE: "not a lot"

WEB ADDRESS:
snuggletiteslumberbag.com

She had tried everything to keep the blanket on and nothing had worked.

Parents, you know the drill. Baby gets bathed, rocked, read to, kissed good night, and tucked in ever so gently. Mom and Dad then quietly creep out of the room and breathe a sigh of relief. Moments later, they hear the dreaded wail: "WAAH!!!!" What went wrong?

Sometimes, it's a "cover up." Or more specifically, the baby has kicked off the covers and is now cold. Why babies do this is a never-to-be-solved mystery. But is there a satisfactory solution to this baby blanket bad behavior? Grandmother Terri Cook says yes.

"The idea for SnuggleTite Slumber Bag first came to me in October 2002. I was talking with my neighbor, Laura, who had a 9-month-old baby and said, 'Surely your baby must be sleeping through the night by now.'

"Laura said, 'No, the baby kicks off the blanket, and then cries because she's cold.' She had tried everything to keep the blanket on and nothing had worked."

Terri was troubled by her neighbor's plight, but after hearing the same complaint from her daughter, Cheryl, Terri knew something had to be done.

"I'm a terrible mom," Cheryl said after she found her son, Jaden, then

6 months old, "shivering and freezing, with his lips turning blue." He had kicked off his blankets on a cold winter night.

Determined, Terri took to her sewing machine and devised a wearable blanket with two straps that hooked on to the baby with Velcro.

A test-run on baby Jaden proved that Terri, who owns and operates a day care center, had more work to do. "Jaden could undo the Velcro, and the straps were too bulky."

So Terri bought new material, devised a baby-proof clip system, and voilà! "Jaden is 4 years old now and he used his SnuggleTite Slumber Bag until he was almost 3."

Cheryl pushed her mom to offer her sleep solution to other exhaustd parents, saying, "Mom, you should put this on the market." Terri protested. "But I'm not a businessperson. I don't know how to do this. And it would be expensive. But Cheryl said, 'If you don't do it, someone else will.'"

So once again Terri got to work, looking for a designer to help her develop the product. "When it came to having the slumber bag and packaging professionally designed, I chose Paul Lapidus at New Funtiers in Palo Cedro, California, to help. I was in unfamiliar territory with this project. Fortunately, I found a great company and person to work with. Without Paul Lapidus and his product knowledge, design talents, experience in the baby product field, patience, and guidance, the SnuggleTite Slumber Bag wouldn't have developed into the product it is today."

Satisfied with the design, Terri wasn't one to snooze on child safety. She had the Slumber Bag safety-tested to make sure it met all federal safety guidelines. Then she found a manufacturer in China and set up a website.

"Everyone loves it," she says. "I've gotten great feedback." It was even applauded by the Canadian SIDS Foundation.

Although putting a product on the market is expensive and building awareness difficult, Terri is optimistic about the future of the SnuggleTite Slumber Bag. In fact, it's become her mission. Until every baby is sleeping through the night in a SnuggleTite Slumber Bag, Terri Cook just won't rest.

Sleep right with SnuggleTite.

Flipple
SpiderStick
PurrFect Opener
CordWrapUp
CDwalltile
Skamper-Ramp
Tilt-A-Fan
Holy Cow Cleaner
Key-P-Out
Mailbox Post Saver
Mop Flops
My Tuffet
Lazy S. Lazy Susans
Appliance Slide
Gel-eez Wristrests and Mousepads
& Golden Grips Cutlery
WordLock

At Home

Author Ann Douglas has said, "Home is an invention on which no one has improved."

That might be true, but many garage inventors have spent lots of money and even more time trying to improve our homes.

This final group of inventors is all about making our home life better. From killing spiders to mopping the kitchen floor while dancing, these inventors hit us where we live. One inventor has a method of turning our CD collection into an art collection. Another wants to wrestle all of those unruly electrical cords dangling from our appliances. What about one cleaner that will work everywhere in your home? One California inventor believes she has the safe solution. What about moving your washer and dryer so you can clean under them? A North Carolina sales executive weighs in with a moving gadget.

From devising a ramp to save animals trapped in the family pool to a gadget to protect your mailbox from vandalism, these eager inventors have slid into home to score big.

Flipple
Turn a water bottle into a baby bottle

A hot summer day, a boat trip, and a fussy baby were the elements that came together to create Flipple.

STAT BAR

PATENT: pending

PRODUCT PRICE: $4.99

STATE: Lives in Illinois, most of work done in Missouri

INVENTOR'S AGE: 52

INVENTOR'S PROFESSION: Retired buyer–full time entrepreneur

MONEY SPENT: $100,000

MONEY MADE: "Very little, but just ramping up now."

Web address: www.babyflipple.com

In August of 2007, Linda Lewis of Galesburg, Illinois, went on a boat trip with her three children and her 9-month old grandson, Jordan. It should have been just another lovely day on her boat. It wasn't. About an hour away from shore, they realized they left the baby's bottle back on the dock.

Panic struck. Everyone on the boat knew they had just a few minutes before Jordan transformed from a sweet sea-loving baby to cranky crazed sea monster.

On board, Linda did have formula and water bottles, so they added formula to the water bottles and let the baby try to drink from a water bottle. What a mess—formula and water everywhere. That's when the Flipple concept was born. "We were passing around a soggy baby all day

and it was then we started discussing the concept of adding a nipple to a water bottle."

Once back home, Linda began working on turning her idea into a reality.

"I began sawing apart baby bottles and buffing the tops off soda tops. With a lot of super glue and crafting glue, I was able to put together a functioning

Flipple packing station

to get Flipple to the point of sale and elevated me to the next level. This came with great anxiety because I had no idea what I was doing."

Linda's first attempt at tooling her product did not go well. The Flipple came back with horrible black streaks that were not baby friendly. The tooling company tried to blame the manufacturing company, but Linda didn't buy their story. Using her work background, she was able to get to the bottom of the problem and had it corrected—at no charge.

Linda has since produced about 12,000 Flipples.

"They were in our front room, our bedroom, dining room, kitchen, and everywhere in our house."

She describes the Flipple as "an innovative new product that turns almost all water bottles into a baby's bottle." When screwed onto the water bottle in one direction, the Flipple is a funnel that allows formula to be added to the small opening of the

water bottle. It can then be unscrewed, flipped, and reapplied. Now a nipple can now be added, turning the water bottle into a baby bottle.

Despite the bumpy start, Linda sees Flipple as her newborn, just starting out in the world and needing the care of its mother. She is not about to give up on her baby.

Linda says she could never have pursued this dream without the support of her husband, Niels. She will soon start to spread the word about her product through ads and through a public relations push. She's hoping mothers will ultimately hear about the Flipple and they will flip over it.

adaptor that was really crude but offered promise. No one would consider using this garage-made prototype on a baby, and the cost of making a professional prototype made me quickly realize this project was not going anywhere."

The Flipple story could have ended here—but it didn't.

"At this point I decided to go all in. I borrowed all the money I could from my 401K and sold everything I owned. Fortunately for me, my wonderful boyfriend, Niels, became my wonderful husband and my house and car were sold for the Flipple cause. All this allowed me

SpiderStick™

Spiderman?

Most of us see a spider on the wall and think, "Let me get a wad of toilet paper and kill it." Chris Crowley saw a spider on the wall and thought he might have a way of turning spiders into cash.

STAT BAR

PATENT: provisional

PRODUCT PRICE: $9.95 (25 in pkg.)

STATE: Colorado

INVENTOR'S AGE: 40

INVENTOR'S PROFESSION: mechanical engineer in medical device field

MONEY SPENT: $600

MONEY MADE: not yet manufactured

WEB ADDRESS: spiderstick.com

"I look at marketing as a big freaking voodoo."

Spiders used to bug Chris. He was like the rest of us. If he saw a spider, he would head to the bathroom, grab some toilet paper, overpower the spider, and flush the remains down the toilet. Some spider safaris don't go smoothly. Spiders have been known to hop out of the toilet paper and escape capture before the final flush. Then there's also the problem of a juicy spider. They can leave a real nasty mess on your wall. Sure, the spider is gone, but the memory—and the stain—can remain for years.

One day, it hit Chris that spiders are a universal problem. Nobody wants those uninvited multilegged houseguests in their home. He concluded that if he

could come up with a spider solution, then he would be the man, Spiderman, a hero—a rich hero.

Chris sat down at the drawing board and came up with the SpiderStick, the ultimate tool to stick it to spiders everywhere.

Here's how a SpiderStick works. When you see a spider, centipede, cricket—you get the idea—you grab a SpiderStick and peel off the backing at one end of the 12-inch-long stick. Doing that exposes a special adhesive. You use that end to slap a surface and "grab" a bug. No more catching critters in tissues. No more spider splatter on the wall. No more running for the vacuum cleaner to scoop up dead bugs. And SpiderStick doesn't leave a gooey residue on the floor or wall. Just toss the stick, with insect attached, into the trash, and you'll never be bugged by that bug again.

Chris's wife, Jennifer, might call

Never be bugged by bugs again

SpiderStick just another of her husband's crazy projects around their Golden, Colorado, home. Over the years, Chris has invented household helpers such as a way for shelves to slide in and out of cupboards, and has crafted a raised deck at the back of their home. But unlike many of the other inventors profiled in this book, Chris is an inventor by profession. In his career, he's invented and taken to market seventeen medical devices through various companies and his own company called Table Mountain Innovations.™ When he got laid off his job as a mechanical engineer at General Electric about a year ago, he wasn't worried. He already had the development of several devices in the works—including a sophisticated pager-like device that signals surgeons when their surgical gloves get a tear during an operation. "If they accidentally get blood on their hands and contract a disease, worker's compen-

sation would have to cover the loss at great expense. I can see a time down the road when insurance companies could demand surgeons use a device like this."

Chris networks around town with other professionals who used to be part of the vendor/employee circle at GE. Now, they still work together as needed—just in different geographic offices. Most, like Chris, work at home and don't miss for a minute commuting in Denver traffic.

Going from highly sophisticated to very simple inventions, Chris's new SpiderStick was actually tagged as "too simple" by the judges of the *American Inventor* TV show. Still, he knows by hard-won experience that a lot of engineering goes into the simplest products.

He currently has a provisional patent for SpiderStick and plans to pay the big bucks to get a utility patent soon. That is, as soon as he finalizes the kind of

adhesive he'll use on the stick. "It has to be gooey enough, yet soft enough, and never leave a mark." Then he needs $3,000 to $5,000 to manufacture the first run. He'll package them as twenty-five sticks for $9.95. While filling the dozens of preorders received via the SpiderStick website, he'll pursue other marketing avenues. "I look at marketing as a big freaking voodoo—it's not scientific at all. It takes a lot of work and tenacity and money. You don't want to lose the family farm over it."

That's why Chris treats SpiderStick more like a hobby than anything else, putting in about five hours a week. In the meantime, he keeps his invention company cooking while working all the bugs out of his SpiderStick dream.

Chris is certain when he brings his SpiderSticks to market, everyone will love them—except spiders.

PurrFect Opener™

Open to a New Opener?

The perfect opener for this story would have involved a sealed bottle, a close-minded homemaker, and an anxious cat. I guess I don't have the perfect opener, but I do have the PurrFect Opener. It looks like a cat, but its mission is to open medicine bottles.

STAT BAR

PATENT: US #D492557, utility #7028359, TM on name and shape of product, © as 3-D piece of art

PRODUCT PRICE: under $10.95

STATE: Michigan

INVENTOR'S AGE: 35

INVENTOR'S PROFESSION: entrepreneur

MONEY SPENT: start-up cost $100,000

MONEY MADE: first year, $1,200

WEB ADDRESS: bamaze.com

"I'm really blessed to have so much assistance through family and friends."

Bob Mazur invented the PurrFect Opener because he was the perfect grandson. Well, at least that's what his grandma must have thought. You see, Bob's grandmother had a mountain of medicine bottles on her table and was unable to open them. "We visited her every Sunday to help with lawn care and snow removal in the winter—a necessity in Michigan. I noticed that she was complaining about medicine bottles and packaging in general. My brothers and I started throwing ideas around for a bottle opener and made a crude prototype from modeling clay to get a feel for the shape and size. We wanted to come up with a design that would help her open aspirin bottles and push-and-twist safety caps.

"My brother, Michael, and I are engineers, and our other brother, Gerard, is a sales and marketing guy, so we guessed that if Grandma needed help with this, others did, too."

To see if there were others like their grandma, the Mazur boys conducted an informal survey at their church. They asked not only if an opener was needed but also what other features were desirable.

Cat got your pills?

Push down and turn.

They took the information they gathered and incorporated that wish list into their design.

Safer, easier, and quicker than a knife, their PurrFect Opener looks like a cat and it can cat-egorically open just about anything. It has a soft underbelly for opening push-and-twist safety caps. The cat's head can be used to pry open aligned-arrow medicine caps. The cat's ears can be used to pierce the foil. The tail is perfect for removing the cotton out of bottles and it can help open soup can pull-tabs. A little hook on the belly will help grab a pull-tab. There's a wedge on the cat's back for splitting pills. And when you need to pop a pill out of a single-dose blister pack, you can push the pill into one of four pockets in the back,

so it doesn't end up rolling under the refrigerator. This cat really knows how to multi-task. In fact, the cat has a built-in magnet so he can hang out on your refrigerator door. That darn cat is so cute.

Bob worked on the PurrFect Opener for five years before bringing it to market. "Dad is an engineer, so he helped. Because Grandma passed away, Mom helped with testing. She still uses the early prototype. It looks exactly like the current product, but it's painted by hand. My sisters Geralynn, Cathy, and Michele help with trade shows, packaging, and mailing. I'm really blessed to have so much assistance through family and friends."

Bob's business plan won several awards, including the "Dare to Dream" student business grant by the Zell Lurie Institute for Entrepreneurial Studies at the University of Michigan. That gave him grant money and office space to help launch the company. Bob started a company he called B.A. Maze, Inc. (as in, Be Amazing). A year after starting production, Matt Shankin joined the company. Today, they've sold enough to work at it full time, putting most of what they make back in the company. "We're spending a lot on advertising, but we're cash flow positive. Our target for 2006 is $750,000."

They sell the PurrFect Opener through their website, catalogs, and Midwest retailers including Meijrs,

Michigan Walgreens, and SavMor drug stores. They have appeared on QVC and are looking at doing more TV. "We're trying to get into big-box retailers, and doing what we can to build a brand and develop new products."

As for the name, our inventor met with a consumer product guru and together they came up with it. Bob explains, "Cats are welcomed in the home, just like we want our product to be welcomed. And cats are independent, like the independence our product promotes."

But for those who prefer dogs, Bob's made the DogGone Opener™. It has the same features, only it's in the shape of a dog.

Ultimately, it's all about lending a helping hand (or paw) so seniors take their medications. After all, a missed dosage could be cat-astrophic and that's why the PurrFect Opener is a perfect solution.

CordWrapUp™

It's a Wrap

STAT BAR

PATENT: design US #D488372-S design (Apr. 2004); utility US #6802471-B1 (Oct. 2004)

PRODUCT PRICE: $0.69–$1.69, depending on size

STATE: Florida

INVENTOR'S AGE: 67

INVENTOR'S PROFESSION: retired electrical engineer

MONEY SPENT: $13,000

MONEY MADE: nothing

WEB ADDRESS: cordwrapup.com

So Michael drew his idea on a napkin.

Cords. They're everywhere. Hair blowers, phone chargers, power tools, oh my! Much of our lives are a tangled mass of cords.

Michael Gambrell, a 67-year-old retired electrical engineer, has strangled his last tool, hanging up that old method in favor of his invention, the CordWrapUp. At one time he, too, had cords wound around every tool in his woodworking shop, so Michael started making wood paddles and wrapping the cords around them. "They worked, but when I used the tool, I took the cord off the paddle and left the paddle sitting on a table far from where I used the tool. The paddle needed to be attached to the cord somehow."

One day, after working the kinks out of some prototypes, Michael brought a few into the house to show his wife, Betty Ann. Her response? "Let me see those." And before you could even say "CordWrapUp," she had her hair dryer and curling iron cords all wrapped up.

That struck a cord. Michael could see how his invention might end cord clutter and also bring in a few bucks. Thanks to Betty Ann, he discovered no less than twenty-five uses in their own house: extension cords, telephone cords, computer cords, the mixer, to name a few.

Michael's son-in-law, Steven Stevilla, suggested he use a mold to make his paddles out of plastic. Michael found a local company to make an initial mold, and filled it with two-part epoxy. The result was hard to use because the plastic wasn't flexible; it got so hard that he couldn't get the cord off! He tried other materials and talked with people, asking how soft plastic items were mass-produced. That's how he discovered the magic of injection molding. He found a local mold-making company through his phone book, but its pricing was prohibitive. The CordWrapUp's run seemed to have wrapped up.

Then, on a cruise, Michael bumped into a friend of a friend, Richard Johnson, who happened to be a patent agent, and Michael learned about the patent process. It could take about three years and cost $7,000 to get a patent. Michael decided right then and there his invention wasn't worth it, but Richard wasn't so sure. So Michael drew his idea on a napkin. Doesn't that sound familiar? Richard looked at the drawing and agreed—the CordWrapUp is such a simple device—it probably wouldn't pay to patent it.

But home on dry land, Richard reconsidered. He looked around, noticed how many cords he had wrapped around things, and saw the potential. He did a patent search for Michael at half price. The only patent close to it was from the early 1920s for something that was molded to a cord. With Richard's help, Michael pursued a design patent, and received it almost one year to the day after applying. Next, he needed the

Wrapped and ready

utility patent, which was also approved in one year. Richard was amazed, because it's unusual for patents to wrap up that quickly.

With his patents in hand, Michael decided to pursue it further. When a Florida injection mold company in Florida priced a mold at $15,000—too steep for this retiree's fixed income—Michael was ready to tie up the invention business. Then someone suggested a guy in Tampa who makes molds in his garage. Michael contacted him, and the guy produced a mold with a lot of flexibility at a reduced price.

Using that mold, a plastic manufacturer in Ormond Beach made 1,000 CordWrapUps for him. He's given them away as gifts and taken them to the Yankee Invention Expo, where he met people interested in the CordWrapUp's future. He's sold some of them, but still has several hundred in his garage.

Before deciding to pay for an injection mold, Michael did some marketing math. He took the population of the U.S.—then over 291,000,000—and divided it by four to get the number of households. He reasoned that if only 5 percent of the households in the U.S. each had only ten uses, not his twenty-five, the marketability of his invention would still be astronomical.

The challenge is: How do you get the word out to the buying public? "My CordWrapUp needs explaining," Michael knows. "And a website is not the answer to making people aware."

Michael plans to continue going to invention trade shows, approaching companies that might be interested, and looking into selling his CordWrapUps in catalogs. Michael is far from finished, so this story is definitely not a wrap.

CDwalltile™

See CDs

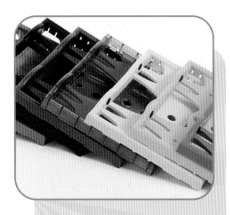

STAT BAR

PATENT: pending

PRODUCT PRICE: $19.95 (9 in pkg.)

STATE: Colorado

INVENTOR'S AGE: 48

INVENTOR'S PROFESSION:
business manager, entrepreneur

MONEY SPENT: $250,000

MONEY MADE: wouldn't say,
licensing potential in the millions

WEB ADDRESS: CDwalltile.com

"I went through eight prototypes before I was happy."

As a teenager, Dean Fallanger worked in a record store. He recalls, "I loved the Beatles, and I especially loved the art on their albums. But in those days, you couldn't stick up an album cover on the wall like artwork. It was too large and heavy and, besides, the colors would fade."

The idea was shelved—until 1992, when CDs became the way people bought music. And the artwork kept getting even better. "I wanted to put CD covers on display in their jewel cases so I could enjoy the art . . . and find my favorites easily."

More than a decade passed while this audiophile pursued a career in computer software and managed a company that processed bone material for dental grafts. A busy man, Dean still found time to sketch his CD display solution: colorful frames for CD jewel cases, which his wife, Doreen, named CDwalltiles. Placed in these plastic frames, the CDs would be stored while the pattern of framed CD cases would become artwork on the wall. The frames could be arranged in a variety of ways, giving the user even more creative freedom.

Starting full-time in November 2004, Dean worked with designers to turn his sketches into plaster-of-Paris–type molds and he kept perfecting—really perfecting—his prototype. "Well, maybe I overengineered it, because I went through eight prototypes before I was happy." Then he selected a Longmont, Colorado, firm called Altratek Plastics, Inc. to educate him on how to take his sketches and models and turn them into a prototype he could test. "Dan Kaven, the owner, was a true blessing in this process and continues to be, as we manufacture the CDwalltile."

Dean also started working with a Denver graphics communication firm called Liquid Inc. to develop the right

Artful storage

look and feel for his logo and packaging. With each step, he was framing his idea to get it ready for display—and for the marketplace.

A month after Dean turned into a full-time entrepreneur, he was scouring retail stores to find a slick way to attach the tiles to walls. Serendipity stepped in. Dean saw a Christmas light display hung on a wall with a new 3M product called Wall Command System. This new product used a removable adhesive. No nails, no holes. Dean was hooked.

"I didn't want people using nails or screws to put up these frames for CD jewel cases because they'd need lots of them to put up a multiple-frame display." He bought the 3M system clips, which could be pulled off the wall, no fuss, no mess, and tested them. Perfect. "I thought, worst-case scenario, I'd have to buy these clips retail and put them in my packages."

To test for durability, Dean and Dan used the clip to hang a CDwalltile on a personnel door at Dan's plant. People opened and closed the door dozens of times a day, through all seasons. The tile never fell—not once.

So Dean went to 3M for permission to use the Wall Command System clips, but 3M made him an offer, a volume purchase price that was a fraction of the retail cost. They also agreed to the use of the 3M logo on the packaging as an endorsement. "Having the 3M logo gives me instant credibility with big retailers. And it also gives me confidence. Once I got 3M's blessing, I knew my product was great." By agreement, Dean meets annually with 3M's VP of Home Division to get approval on packaging and anything else that represents 3M's brand.

Serendipity struck again early in 2005. Dean had placed a $250 ad on the ascensionshowcase.com's art and unique furnishings website, thinking, "It can't hurt." That ad generated a call from a distribution company in Europe wanting exclusive European rights to CDwalltile.

"I did some research and learned that 780 million CDs are sold in the U.S.

each year—and European CD sales per capita are even higher than here." Putting together details of the European distributor's contract has kept Dean hopping, "often fifteen hours a day." That includes scheduling a special run of 20,000 nine-packs and printing exclusive packaging for the European market. This could lead to millions in sales a year—and that's even before he completes pending agreements for Asian and Australian rights. When it comes to manufacturing, assembly, and fulfillment, Dean has kept the business local. In addition to having Dan's company manufacture the frames and doing bulk order fulfillments, he works with Foothills Gateway in Fort Collins. This company employs mentally disabled adults to assemble pieces for all kinds of businesses in the area.

To market in the U.S., Dean is now seeking experienced salespeople who have existing contacts in retail stores, and he plans to hit freshman college students with an impressive ad to draw potential buyers to his website. After two years of setup work and impressive initial success, Dean says, "I had blind faith when I started this. Thank goodness my wife, Doreen, and our two sons have been supportive."

Now surrounded by 300 CDwalltile frames in colors chosen from his palette of thirteen choices, this inventor has turned his love of music into a beautiful sight.

Skamper-Ramp®

Ramp to the Rescue

STAT BAR

PATENT: US #664389

PRODUCT PRICE: $49.95

STATE: South Carolina

INVENTOR'S AGE: 63

INVENTOR'S PROFESSION:
artist; president of sign
manufacturing company

MONEY SPENT: $4,000 in legal fees

MONEY MADE: just became
profitable, "better than a wash"

WEB ADDRESS: skamper-ramp.com

"The quick turnaround on getting the patent is what patent lawyers call a hole-in-one."

Tom Davis's three grown kids and five grandkids love to gather at the Davis family backyard swimming pool for some cool, wet relief from the hot, muggy Charleston, South Carolina, summers. But what Tom, an animal lover who has two cats (Phoebe and Gypsy), two rats (Cheddar and Cocoa), and a baby squirrel (Squire) didn't love was what he calls the "dirty little secret" of home swimming pools.

That dirty little secret is the high number of animals that drown in the pool on a regular basis. "Just ask pool maintenance people. They'll tell you. I just got tired and really sad with all the little critters I would find dead in my skimmer basket," says Tom.

As a nonpool owner, that was a secret I had never known.

At 3 a.m. one night a few years ago, Tom's dog, Brigette, barked. The bark alerted Tom to the presence of "the biggest momma possum I've ever seen." The possum was in trouble, because despite the fact that she could swim quite well, she couldn't find her way out of the pool. Without Brigette's barking and Tom's lifesaving rescue, the momma possum surely would have drowned.

Tom had had enough. "There had to be some way to solve this problem. I had the idea to build a ramp, and that weekend I fooled around with the design until I found a ramp that worked."

Tom found that a 30-degree angle and the color white worked best. And

Ramping up the safety for "critters"

whether it was the family dog, an uninvited frog, or a curious squirrel, the rescue ramp did the trick.

"I used it in my pool for two swimming seasons and I never found another dead critter," he says. People would come by Tom's house, see the device, and comment on what a good idea it was. Which got Tom thinking that perhaps other pool owners, and other little animals, could benefit from the ramp he'd invented.

He got a patent quickly. A year and one patent later, he had the Skamper-Ramp, a strong, lightweight, floatable plastic device that provides animals with an escape route from swimming pools.

"The quick turnaround on getting the patent is what patent lawyers call a hole-in-one," Tom says.

In developing the Skamper-Ramp, Tom networked with friends to find a plastics expert who could help him select the best material to construct his product. That led him to Jay Wadell, a plastics expert and "marketing guy" who not only helped Tom find the right plastic, but who also took a strong interest in the Skamper Ramp. Jay ramped up the business and now Tom has turned over the operations of the company to him. With Jay watching Skamper-Ramp, Tom can get back to

his "regular job" as the owner of a sign business that makes products for interiors of restaurants, including Applebee's and Outback Steakhouse.

Looking back, Tom reflects, "My idea wasn't really that ingenious. I just saw a need and thought, 'This is silly for them to drown.'" For Tom, the best part of inventing the Skamper-Ramp is that it works. "It actually saves little critters' lives."

Tom exposed that dirty little secret and saved a whole bunch of critters in the process.

Tilt-A-Fan™

Updraft

STAT BAR

PATENT: US #6779768

PRODUCT PRICE: $6.99

STATE: New Jersey

INVENTOR'S AGE: 51

INVENTOR'S PROFESSION:
electric utility company dispatcher

MONEY SPENT: $42,000

MONEY MADE: $20,000

WEB ADDRESS: no active website—
"Look for it on eBay."

". . . if it's cheap enough, people will buy it."

Dan Ferre has always appreciated the cool breeze from a fan. I guess you could say he was a big fan of fans. But Dan found a way to make his fans even better.

Dan has always liked to work on cars. He worked as a mechanic for seven years after high school. Even after he found a better job, he still would tinker in the garage. "I have a few antique cars, so I'm in the garage a lot. A fan comes in handy while I'm working at the tool bench in the summer."

Dan used to stand his box fan on the floor, but he got tired of the air just blowing on his legs. To cool the sweat of his brow, he started leaning his box fan against his toolbox. Sure, that angled the air toward his face, but the toolbox cut down the airflow. Not good in a hot sticky New Jersey summer. Dan set the fan on top of his toolbox, but kept knocking it over. "I'd get involved in what I was doing, forget where the fan was, and bump it." He broke a few fans that way.

"I figured I needed to make some kind of a bracket to hold the fan so it would direct the air up." Dan always has metal around his shop, so he made a stand out of metal.

"It worked so well that my wife, Kathy, started using it indoors in rooms that don't have good air circulation, and outdoors on our patio. Friends liked the stand and asked where we bought it. But I had made only one. Since adjustable fans cost $30 or more, they convinced me to make it available to everyone." Dan talked with people in hardware stores to see if the interest was even broader. He learned that "if it's cheap enough, people will buy it." Dan dubbed his invention the Tilt-A-Fan Stand.

"I knew I couldn't sell one out of metal, in case, by freak chance, it would

Working all the angles

break. A metal rod going into a motor could cause an electrical short." So Dan went full tilt researching plastic. He started with the material plastic milk cartons are made of. It was too flexible. Then he tried ABS Plastic, but after ten flexes, it would break. It was too brittle. "The guy who makes it for me suggested Super Tough Nylon 66. After 100 bends, it was still flexible." Super Tough was just right.

Manufacturing became a costly process. Dan found out that molds are very expensive. And hiring a top-notch patent attorney is expensive, too, because that person has to be so thorough. "A friend of mine got a patent years ago when he made a tool. He thought it would be a good idea for me to get one. He told me that a big company could take my idea and sell it for less, putting

me out of business. With a patent, at least I'd be protected."

It took about a year to get his patent, and another year to get the Tilt-A-Fan made, trying out various plastics and testing it. He tested it on different flooring, such as linoleum, tile, and carpet. He'd put it on and let the fan run for hours. The design took a little tweaking here and there, and a good two years before one was ready to go out.

"When I first got the patent, I was flooded with letters from marketing companies. The most persistent one wanted $25,000. I would get to keep all the sales profits. If I dropped down to $20,000, they wanted 3 to 5 percent of my sales; if I paid them $15,000, they wanted a higher percentage than that. I said it was too much; I couldn't afford that."

His marketing plan was to send a sample of the Tilt-A-Fan to the corporate headquarters of all the big companies. "That didn't work well because of the volume of the requests they get. Most said, 'No thank you. We don't know about this product.' But they had it in their hands! What they needed was the endorsement of a marketing company saying, 'Hey, you need this.'"

Dan got his Tilt-A-Fan into some catalogs, but wasn't selling as many as

he'd like. His cousin told him about a friend who worked in marketing. "Suzanne is helping me now. I pay her a percentage of sales, no up-front money. She's doing the same thing as a marketing company, talking to trade shows and big companies about my product. She was impressed I got it into three catalogs on my own. Two venues, the Harriet Carter catalog and *Healthy Living* magazine, said 'no thank you' to me three years ago, but through her, I got in them both."

Dan knew the marketing was starting to work when family and friends started to see his gadget for sale. "My sister, Deanna, called me yesterday, 'Oh my God, you're on page 50 of Harriet Carter.'"

It nearly blew her away. And when you're selling a fan accessory, that's a good thing.

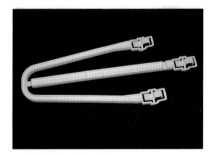

Flat and easy to store

Holy Cow Cleaner™

Other Cleaners Should Moo-ve Over

STAT BAR

PATENT: information not provided, name trademarked

PRODUCT PRICE: $2.99–$3.99/bottle

STATE: California

INVENTOR'S AGE: 49

INVENTOR'S PROFESSION: writer

MONEY SPENT: $300,000+

MONEY MADE: undisclosed

WEB ADDRESS: holycowproducts.com

"It's as if these folks have almost had a religious experience."

Imagine a world where there was just one cleaner. But this cleaner could clean wood, tile, carpet, dishes, laundry, metal, plastic, marble, granite, sinks, tubs, oven, soap scum, furniture, terrazzo, stainless steel—everything. Just think of all the extra space we'd have under our kitchen sinks!

Joni Hilton had written sixteen books. In one, *Housekeeping Secrets My Mother Never Taught Me,* she recommends all sorts of cleaning products for all sorts of jobs. While writing that section of her book, she kept wondering about finding one product to replace the army of bottles under her sink. Joni says she's not a neat freak, but she does like to have a clean home. And with three sons, a daughter, and two dogs, she knows how tough it can be to battle dirt.

You may know Joni's husband. If not his face, his voice. He's the Bob in "Bob, tell them what they've won." Bob Hilton has hosted a few game shows and been the announcer for many more (including *Let's Make a Deal, Truth or Consequences, $10,000 Pyramid, Newlywed Game, Dating Game,* and *Joker's Wild*). He's also hosted some talk shows and anchored the news.

Bob didn't share Joni's passion for finding the perfect cleaner, but that didn't stop her. She approached chemists about developing one cleaner that was safe and could clean anything. Joni believed if she found it, she could clean up.

"I had to beat all the other products by a huge dramatic margin, or how else could I beat a mega company like Proctor & Gamble?"

Most of the chemists just laughed, but she finally found one who thought

Keep it clean.

he might be able to do it. Two years later, she had a prototype. In the lab, the chemist dripped competing cleaners, plus her prototype, down a tile covered with train engine grease. "Most of the others didn't budge the grease, and a couple of them left a gray streak where you could see they were working a little bit. Oh—and these others came in concentrated form. Then he dripped our diluted product down the tile and it looked as if we were using white shoe polish. It stripped the tile to pure white instantly."

Joni was amazed. But that wasn't enough. She wanted a safe product. When the chemist told her it didn't contain one harmful chemical, Joni's jaw dropped. This was the breakthrough product she wanted. When Joni's husband saw how it worked, he jumped on board.

Next, this miracle product needed a name. Joni said when she first saw how it cleaned she'd said something like "Holy cow!" Not exactly, but since the product is a cleaner, she cleaned up the name.

A friend of Joni's made an "udderly" adorable bovine logo. Joni and Bob filed the legal paperwork, hired brokers and a national sales director, set up an office, and began showing store managers what Holy Cow could do.

From the moment customers started trying Holy Cow, Joni says, the feedback was amazing. "Our website features unsolicited customer testimonials that are unbelievable. It's as if these folks have almost had a religious experience. But I understand, because I shared their same frustrations for years with expensive, caustic products that didn't work as promised. To find one that's this remarkable is very exciting." Even better, "And to think that if people would just switch to Holy Cow they could toss out all their toxic cleaners and never worry again about toddlers getting into poisons. Holy Cow can literally change the way America cleans."

Holy Cow is now found all over California and Arizona, and its distribution is slowly creeping east. Joni says this has been a huge learning experience. "I can't believe how much I've learned about trucking, palettes, purchase orders, warehouses, supermarkets, and shelving fees. Did you know manufacturers pay for shelf space in supermarkets?"

Finally, Joni told me about a Denver man who, a few months ago, had tried to poison his wife by substituting Holy Cow for her mouthwash. It triggered police and hospital reports, and poison control people contacted Joni at Holy Cow. She assured them there was no cause for concern. Obviously, this Denver husband hadn't read the front of the bottle, which clearly states: "WORLD'S SAFEST CLEANER."

So not only can this cleaner clean every surface and do it safely—it also prevented a murder. Holy Cow!

Key-P-Out™

No Turnkey Operation Here

STAT BAR

PATENT: pending

PRODUCT PRICE: $19.95

STATE: Florida

INVENTORS' AGES: both 40

INVENTORS' PROFESSIONS: Joe, sales manager; Ron, operations, manager

MONEY SPENT: $80,000–$100,000

MONEY MADE: $5,000
(sold at wholesale)

WEB ADDRESS: key-p-out.com

. . . she knew others had keys to her place, so she slept with a knife under her pillow.

When he worked the night shift, Joe Lubrant slept most of the day, and learned firsthand that apartment maintenance people can just let themselves into your apartment, turning the key halfway as they knock.

They may have the legal right to enter, but it's an uncomfortable feeling having someone just waltz into your home. "I started to realize that even though my door was locked, I had no privacy and, more importantly, I had no security. I wondered if the locks had ever been changed between tenants. Who beside the maintenance man had a key to my place?"

Most leases don't allow you to modify the lock, but Joe was intent on finding a way to keep people out until he let them in. Some years later, Joe met Ron Kent, who was also looking to start a business. They knocked around some ideas, and Joe mentioned his about keeping people out. Ron got keyed up—in a good way. They got to work on a widget that keeps a lock intact and doesn't require installation.

"We made fifty different drawings," Ron says, "but applying them to a product is a different story." Joe adds, "We started working with steel, because when you think of locks, you think of steel. But we quickly realized this was a leverage issue and set out to make plastic prototypes." This allowed color options and avoided scratched-up doors from the steel prototypes. After six design changes to the handmade prototypes, they committed $400 to have a local machinist make one out of a solid plastic sheet. "Then we had a CAD drawing made so we could move on to the mold-manufacturing search and submit our patent request."

Next, Joe and Ron went looking for a manufacturer. They started with the phone book, but ended up with a friend of a friend who connected them to a manufacturer in the East. They sent their CAD drawings, and got a quote for a mold and production run.

Soon they had 5,000 in three different colors sitting in a warehouse ready to go. "Then we got the news that what we thought couldn't be improved upon, could," says Joe. They had given out hundreds to a marketing company that tested Key-P-Out all over the country. With hurricane winds so prevalent in south Florida, building codes there require that doors open out, not in, to an apartment. That meant smaller doorjambs. Joe relates, "That tightened up the area we had to work with, because the dog bone style of the Key-P-Out was rubbing up against the doorjamb. We had to get rid of the bottom loop so it would fit smaller doorjambs. We cut one up, got rid of the loop, and noticed it looked like a key." Isn't that fitting—a Key-P-Out that looks like a key.

They had another mold made. "We weren't exactly happy to have to modify it—it required new drawings and a new mold. But using the key shape made it worthwhile. The bone shape would fit 90 percent of doors, but when we're paying for a TV campaign, we wanted it to be 99.9 percent effective."

That TV campaign came about when a friend of a friend hooked them up with

their marketing godsend. The new addition to the team had experience selling items on TV and the Internet. They created Secure America™ to promote Key-P-Out, using 15- and 30-second commercials. Secure America has the exclusive rights to sell the product while Joe and Ron maintain the manufacturing and the patent rights. The agreement guarantees the sale of 1.5 million units in three years, or else Secure America loses its rights. Sounds like secure deal.

"While meeting with Secure America reps in Orlando to lock in the deal, we gave a Key-P-Out to our waitress. She told us she had been scared to death in her apartment because she knew others had keys to her place, so she slept with a knife under her pillow. The people we were meeting with to convince them there was a market for the Key-P-Out thought we'd set this up!"

Joe and Ron have done quite well selling the Key-P-Out through their website, and mail and phone orders, and getting newspaper publicity through newspapers before closing down their selling efforts in favor of transitioning over to Secure America and its exclusive deal. They've spent five years and, according to their accountant, $80,000 to $100,000 (more in pain and suffering, according to our guys) just to get this far. But now, it seems the door will soon be swinging wide.

Looks like a key but works like a lock.

"It's been exciting," says Joe, "but we've had a lot of fun with it. How many people have actually sold an idea they've come up with?" And now that they're keyed in, "However well the Key-P-Out does, we've agreed we'll do another product. We've learned a lot." Ron adds, "Still, I'm not confident that this is the one I'll use to write a book about how I made my first million."

And maybe not being overly confident will be the key to their success.

Mailbox Post Saver™

Mailbox Baseball Called on Account of Brains

". . . residents on my route end up replacing their mailboxes as frequently as seven times a year."

When you drive a school bus for ten years, you really get to know a neighborhood. You notice when a house gets painted, when the bushes are trimmed, when someone gets a new car. As a bus driver on the rural roads around southern Missouri, Jay Sutton saw changes in his neighborhood and didn't like what he saw.

Jay kept seeing more and more smashed and damaged mailbox posts on his route. Snowplows cleaning the roads in winter mangled some mailboxes, but most were vandalized by teenagers joyriding in the middle of the night. "They play a game called mailbox baseball, using a baseball bat to swing at mailboxes as they drive by. Completely cutting off a mailbox from its post is their version of a home run."

A little Internet research shows Jay wasn't making this up. Wikipedia says, "Mailbox baseball is the act of using a baseball bat to knock over roadside mailboxes while a passenger in a car. It can either be played as a game, keeping score as in baseball, or played just for fun. Either way, it is considered a form of vandalism." If mailbox baseball is defined on Wikipedia, it must be prevalent in many communities across the country.

Jay adds, "The fine for getting caught at this game is $50,000, but it's rarely enforced. Consequently, residents on my route end up replacing their mailboxes as frequently as seven times a year."

One day in 1988, Jay got especially riled. That's when he saw so many smashed posts, he started designing a vandal-resistant one that could be moved ten to twelve feet back off the road. Since then, he's applied his metal-working skills from his days as a punch press operator to forge a steel post for regular-sized mailboxes. The idea is to position the mailbox at the side of the road for about two hours a day (when mail deliveries are due). After retrieving their mail, owners would move the mailbox off the road.

"My current prototype of an extended, movable metal post has to be moved manually. But I'm working on a motorized one driven by a solar battery. It could even be put on a timer."

The perfect prototype has been a long time coming. "I'd build one or two prototypes a year, always reusing materials from the previous one, so it wouldn't cost me a lot in materials."

Jay was proud enough of his latest prototype to apply to appear on the *American Inventor* TV show. "I made it to the second round before I was told my post saver isn't what they're looking for." But he'd still post the experience as a great one, based on the positive feedback he got. As he says, "My family and friends have always been supportive, but families are known to say whatever you want to hear to make you happy. I needed the judges at *American Inventor* and others who don't know me to tell me I had a good idea."

Jay hopes his invention will cause mailbox baseball players to strike out!

With his wife, Lorraine, Jay has always put family first, which hasn't left him a lot of time for building prototypes. "Lorraine and I love kids. We've been foster parents over the years, and we adopted six kids in addition to bringing up our own." Their "kids" now range in age from 4 to 32, plus a 3-year-old grandson who lives with them.

Several years and $3,000 later, Jay has a provisional patent, a prototype that works, and a dream to license his mailbox post saver to a manufacturing company. "This movable post has to be easy and convenient to use. When the mailbox is being moved, it has to stay level so the rain won't get into it. These days, I spend almost four hours a night

on the computer researching motors and gear mechanisms for the motorized version. But I know that making this post on my own would be very expensive. That's why I'd like a manufacturer to step in."

Jay will soon get a chance to display his moving metal post on Australian television. He's optimistic and determined that people won't have to be victimized by baseball-wielding vandals—even if he has to set up a company to make and sell his mailbox post saver himself.

While most of us might see mailbox baseball participants as bat bait, this bus driver is taking the high road, and seems to have come up with a better mailbox.

Now that's a hit.

Mop Flops™

"I'm Not Dancing, I'm Cleaning My Floor"

"It makes keeping my floors clean like a dance in the park . . . "

Some inventions come from chore avoidance. The inventor hates to do a necessary task, so he/she invents a solution. That's the story behind Mop Flops, a hands-free way to clean a floor. You see, this inventor hates cleaning and had her sewing machine set up in the kitchen. Add one part cleaning-phobe to a sewing machine, plus four kids, and you have all the right ingredients for this invention.

Because the last two of Gaile Spalione's four children were born only a year apart, she was always holding one while the other young one was nearby on the kitchen floor. Since the older children were 8 and 5 years old, Gaile was often in the kitchen. Now, you might think that someone who invented a cleaning product would be obsessed with cleaning, but not 46-year-old Gaile. She's no neat freak. Still, it did bother her that her young kids sometimes played on a filthy floor, so she often threw down kitchen towels and shimmied around to give a quick clean while holding the baby.

Then one day, she shimmied by the kitchen table and bumped into the ever-present sewing machine. That's when it

clicked—or flopped—into place. Gaile sewed elastic around the tops of the towels and wore them, never missing a beat in her mopping dance again. When her friends wanted some for themselves, she knew she was onto something.

The current version came about by the urging of her patent attorney. Apparently, you can't patent an elastic-topped towel. So to make the design unique enough to patent, she went back to the sewing machine—using terry cloth instead of tea towels, lining them with vinyl—and designed a Velcro pocket that would hold a sponge. Gaile named her invention Mop Flops.

Here's how Gaile's packaging describes Mop Flops. "I'm a mother of four. My kitchen floor doubles as a speedway for Hot Wheels, a drive-in for Barbie, a spill zone, and, oh yeah, a dog run! My kitchen is a twenty-four-hour restaurant with do-it-yourself service and 'Mommy, can I use the big kid cup?' patrons. By the end of the day, I need a bulldozer. My hands are seldom free . . . so, I created Mop Flops!"

While her friends' support got her out of the kitchen and into the marketplace, it's Gaile's mother who has supported this invention financially. Her mother's belief in her is what keeps her going—you see, she has to pay her back. Now she's added money of her own, and a friend has backed her. Together, they've all invested $75,000 over the last four years.

The first year, she worked with a manufacturer; the next two, she worked on patenting it; and this last year has been about getting the word out. Her advice to a new inventor is to market the product before manufacturing it. She spent all of her money manufacturing, then had no money to market Mop Flops.

"I keep my Mop Flops on the floor, and slip them on and off. I use them every day. It makes keeping my floors clean like a dance in the park instead of a chore. I can walk on a freshly washed, wet floor, and I use them to quickly dry mopped floors so I don't have to wait to

get back in the kitchen. I can't live without them."

Slowly, others are learning that they can't live without Mop Flops. Gaile sells about five pairs a week. While this doesn't put food on the table, at least they're selling, and that's without any advertising. Gaile recently hired a PR agent who sends releases to newspapers and magazines. But the best piece of advertising she's ever enjoyed was a spot on the *Oprah Winfrey Show* for her Million Dollar Idea Challenge.

Gaile had heard about the contest and immediately overnighted her Mop Flops

to *Oprah*. The next day, Harpo Studios invited her to the show. No one was more surprised about Gaile's appearance on *Oprah* than her husband, John. He hadn't been the biggest fan of Mop Flops. And let's face it, the *Oprah* story sounded a little farfetched. But Gaile convinced John that the *Oprah* contest was real and on October 30, 2003, she went on the show.

The *Oprah* appearance helped sales, but Gaile is still looking for an even bigger break to make this mommy-made product really clean up.

My Tuffet™

There's No Business Like Shoe Business

STAT BAR

PATENT: US #6877817

PRODUCT PRICE: $59–$119 and up

STATE: Maine

INVENTOR'S AGE: 55

INVENTOR'S PROFESSION: "domestic engineer"

MONEY SPENT: first year $38,000, and more each year

MONEY MADE: sold more than 1,500 tuffets

WEB ADDRESS: mytuffet.com

. . . when Nancy gave her friends the footstools with feet, they all went crazy.

We know the nursery rhyme "Little Miss Muffet sat on her tuffet eating her curds and whey." But do we all know what a tuffet is? Well, it's a 17th-century word for footstool. Sadly, I didn't know a tuffet from a tiara. I learned the difference from Nancy Brown, who is in the business of making "tuffets with an attitude."

It all started because Nancy is part of a group of nine women who have been getting together for twenty-five years. When their kids were little, they met once a month for lunch. As the kids grew up, the lunches got longer. Every Christmas, the women would swap gifts. A few years ago, Nancy was buying Carlos Santana high heel shoes at Nordstrom when it hit her—not the shoes, but a great gift idea. She decided that since her friends were all crazy about shoes, she would make them gifts that were all about shoes.

Nancy went home, grabbed shoes from her closet, and arranged them into a few groupings. She took three shoes and put them at 120 degrees, making a circle with the three heels at the center and the toes facing out. She put soup cans in the shoes and placed cardboard from pizza boxes on the shoes and then a pillow on top of the cardboard. That was it. She looked at the prototype of her footstool and decided that they would be her holiday gifts to the girls that year.

Then Nancy went to the Salvation Army store and picked up fourteen pairs of high heel shoes. She also got twenty-seven soup cans. She then bought tapestry fabric, some fringe, foam, and ¾-inch plywood cut into 16-inch circles.

She put them all together, and added buttons and black felt underneath. She used closet poles to make ankles (this way they all got thin ankles) and covered the poles with knee-high stockings.

That Christmas, when Nancy gave her friends the footstools with feet, they all went crazy. They also agreed that these footstools were so cute, so unique, that Nancy just had to go into business. Since her children were all out of the house, this Maine homemaker decided it might be something fun to do. She talked to her husband, who encouraged her to take her footstools to the next step.

Nancy went back to the drawing board and re-engineered her footstools to make them stronger. She also tried a few different models, making tuffets from rollerblades, ballet slippers, sneakers, men's shoes, kids' shoes, and golf shoes. You name it; if it was footwear, Nancy made it into a footstool.

At about this time, Nancy's grandfather passed away, leaving Nancy and her

brother some money. He used the money to put a deck on his house; Nancy used it to pay an attorney to complete the patent process.

Since going into business, Nancy has seen her Tuffet sales go up and down. In the first three years, she sold about 700 Tuffets, mostly at craft fairs. Since then, they've taken over her house. Her children's rooms are filled with Tuffets. When I asked Nancy what her plans included, without missing a beat she said, "I want to get someone else to actually make the Tuffets."

Nancy now sells them on her website and at a few posh stores, including one on the Upper East Side of Manhattan. They sell for from $60 to

about $400, for custom footstools. This year she just started to turn a profit. "My initial year I spent $32,000 for patent search, patents and associated fees, material costs, display costs, and booth fees for five shows. Since then I have done thirty-eight shows, sold over 1,500 tuffets, done many custom pieces, and have an active website. The more shows I do and the more upscale, . . . the higher my costs . . . including booth fees, accommodations, meals, parking and GAS! I recently purchased a 6 by 10 utility trailer . . . which translates to 10–11 mpg."

Nancy turned a bunch of old shoes into a footstool business, and she did it by not sitting on her tuffet.

Tuffet invasion

Lazy S. Lazy Susans™
A New Spin on Lazy

STAT BAR

PATENT: information not provided

PRODUCT PRICE: sold 10–20 for $150 each

STATE: New York

INVENTOR'S AGE: 37

INVENTOR'S PROFESSION: industrial designer

MONEY SPENT: $500, plus $2,000 in her time

MONEY MADE: $1,500–$3,000

WEB ADDRESS: elseware.to

"I enjoy being open-minded and inquisitive about how things work, then exploring new ways to create them."

Sometimes inspiration comes from reinventing an old standard. That's the case with Janet Villano and her Lazy Susan. Lazy Susan is not a nickname for Janet's slothful slow-moving sister. In fact, I don't think Janet has a sister named Susan. As I'm sure you know, a Lazy Susan is a rotating serving plate placed usually in the middle of a round dining-room table. The concept has been around for generations; Janet thought it was time for Susan to continue to be lazy, just in a different way.

"There's always an artistic aspect to examining how things work that I find satisfying." Janet is a full-time inventor who constantly brings an artistic bent to invention.

Janet's Lazy S. design for Lazy Susans exposes the inner workings of this revolving plate, but with a twist. Instead of placing the ball bearings that make it revolve in a conventional circle, she spaces them randomly in her Lazy S. designs. The concept that traditional Lazy Susans have ball bearings in a conventional circle, or that they have ball bearings at all, might be news to you. But you'll need to trust me on this one.

"With this random configuration of ball bearings, it looks like the plate simply won't spin. But the ball bearings don't know about being conventional. They just work." Janet knows what she's talking about. After all, she's an award-winning product designer who earned her master's degree in industrial design from Pratt Institute.

Her Lazy S. plates or disks, measuring 12 inches in diameter, have been made to order in different variations and sell for $150 each. "They were designed as

prototypes for a design show, and all of the original pieces were sold soon after. I spent a few months reviewing the feasibility of developing them for mass manufacture. But I didn't have the resources to create the molds that would bring the costs to a reasonable level."

Besides, life is more about the inventing than the manufacturing for this 37-year-old inventor. "I enjoy being open-minded and inquisitive about how things work, then exploring new ways to create them."

Never one to take household objects for granted, Janet could never be called lazy. She's designed a whole shelving system above the stove in the kitchen of her Brooklyn, New York, home. She made it so the knife block sticks out in a practical way—

a fine example of artistry blended with usefulness.

If you're not lazy, you can check out more of her avant-garde designs at www.elseware.to, where you'll see her star stool, vortex lamp, volcano bathroom clock and mirror, pierced bathroom towel rings, shower lamp, and more one-of-a-kind designs. With four other Pratt Institute grads, Janet co-founded the Elseware artists' cooperative in 1999. It features the work of industrial designers who love to "blur the line between art and product." Members of the collective produce gallery shows of their products (including Oliver Beckert's Aquariass and Dan Harper's Ashhole, also featured in this book—pages 128–129 and 204–205 respectively). The active members do design consulting, creative thinking, and problem solving for their clients.

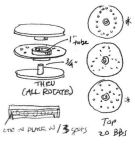

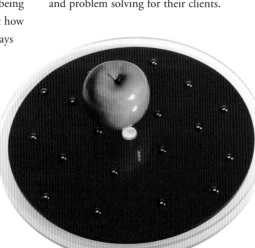

To earn a steady paycheck, Janet does product design for a New York manufacturer of air purifiers, fans, and humidifiers. For now, she won't be applying for a patent for her Lazy S, not out of laziness, but "because I don't want to do the manufacturing and marketing, though it could be interesting to get the design licensed someday."

Janet continues to take ordinary items like furniture and clothing, find the artistry in them, and then reinvent them to be more interesting and functional.

So, sure it might be a Lazy Susan, but it's certainly not a lazy Janet.

Appliance Slide™
Slide, Slide, C'mon, Washer, Slide Home!

Mike Coleman's wife is fanatical about cleaning. Nesha cleans literally everything from top to bottom, with Mike doing the heavy lifting. "I've moved the washer and dryer out a number of times. After tearing the linoleum twice, I told myself, 'This is wrong.'"

Mike put hardwood down, and pads on the feet, but moving the machines put dents in the floor and made the hardwood look bad. He tore it up and put tile down. The heavy machines scratched the tile.

Mike wasn't going to let this slide. He knew this weighty problem had a simple solution. Funded by his uncle Jerry Helms, who said, "If you build it, the money will come," Mike shut himself up in the garage for eighteen months. He set up "a mini-manufacturing site in there. My neighbors went crazy, hearing saws and power screwdrivers. But I wouldn't tell them what I was doing. I'd say 'It's top secret stuff.'" Eventually, Mike came up with a working prototype for the Appliance Slide.

Having been an overseas buyer, Mike asked some top-notch guys in Beijing to build a marketable prototype for him, then he took his idea to the patent office in Charlotte, North Carolina. "My patent attorney said he hadn't seen anything like this before. 'Let's give it a whirl.'" A few months and several thousand dollars later, it looked like Mike had a winner.

The invention is "dumb-ass proof. I extended the floor track four inches, so the person installing it won't jam it against the wall, leaving no room for the hoses. And when you slide a washer or dryer out using the Appliance Slide, it leaves five inches of slack in the hose, so it won't pull the hose out of the wall."

Testing the completed prototype, Mike was moved by the results. "Anyone can move this thing. Even if you're in a wheelchair you can operate it. You can push it in with one finger. Now if your machine goes off balance, instead of those legs digging into the floor or the machine walking across the room, the Slide locks it into place. Instead of eating

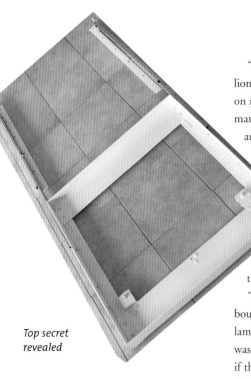

Top secret revealed

up your floor, it does no damage at all."

Mike got his utility patent and put the Slide into production. "Now I'm ready to sell the patent. I've got three other inventions in the works, but I've got to move this product before I do anything else."

While the patent was pending, there was an offer from a large retailer. "These guys offered us $38 million instantly, and they wanted an exclusive. This is where the money becomes mind-boggling. Millions can cloud your decisions very rapidly." Mike took the contract to his attorney, and discovered the fine print wasn't so fine.

"They were going to pay us $7 million up front, then pay the residual based on royalties on the number of units manufactured. But they wouldn't let us audit their manufacturer. They could underreport how many units they were manufacturing. Then we found another paragraph saying that after fifteen years, they had the right to drop us. It wasn't a good situation. We couldn't do the deal.

"After leaving my attorney's office I bought two 40-ounce Millers, sat by the lamppost, and cried. My wife thought I was coming home with a big check. But if they're not on the up-and-up, we can't deal with them." The invention game has its heartaches.

After Mike dried his eyes, he saw the silver lining. He hadn't fully realized the value of his idea before this lucrative offer. "It solidified a figure for what the patent is worth. Marketing is not my forte. I wish it was, but it's not. Without marketing experience, you hit a brick wall pretty quick—regardless of how good a product is. We're experiencing that brick wall now.

"We've spent $300,000 on this, so we're really motivated. We'd sell the patent for $4.2 million. My ideal situation is that, when someone buys it out, I'd be hired as a consultant to work on future projects related to the Appliance Slide."

Mike's friends and family didn't see the Appliance Slide's potential. Even his dad thought his son should let this one slide. A frustrated inventor himself, "He told me, 'I spent a whole lot of money on nothing. You're probably going to run into the same thing.'"

Mike's wife, Nesha, has supported Mike's inventive ways wholeheartedly. "I'd be in the garage until 1:00 or 2:00 a.m.—sometimes I'd work all night. Then I'd come in, get a shower, and go to work."

But Mike read that 16 million washing machines were projected to come into the U.S. market in 2006—up 15 percent over the previous year. "Even if you only sold to 3 percent of the market, that's millions in revenue. And think about the rental market—how many floors get cut up when renters move out of apartments?" Not to mention homebuilders that could supply this slide to buyers right off the bat. "The task has been getting it to the right people to support the patent or buy the patent out."

Mike would like to see an appliance maker put an Appliance Slide on every machine, and ship it directly with the Slide installed. Great news for husbands and wives who are pushovers for clean floors. Eventually, I'm sure Mike Coleman will slide into home with this one.

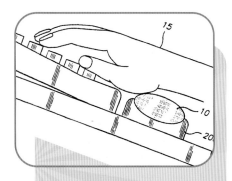

Gel-eez® Wrist Rests and Mousepads / Golden Grips® Cutlery

It's All in the Wrist

STAT BAR

PATENT: US #5566913 (Gel-eez);
US # D494426 (Golden Grips)

PRODUCT PRICE: Golden Grips
Cutlery 24.98–$26.98/knife;
Gel-eez Wrist Rests $6.99–$7.99;
Gel-eez Mousepads $6.99–$8.99

STATE: Washington

INVENTOR'S AGE: 45

INVENTOR'S PROFESSION:
software program manager

MONEY SPENT: Gel $20,000;
Golden Grips $25,000

MONEY MADE: Gel $750,000;
Golden Grips $15,000

WEB ADDRESSES: goldengrips.com &
gel-eez.com

Then David hit on a "brrreakthrough" idea.

David Prokop has kept a lab book cataloging his ideas for twenty years. Without fail, two to three years after coming up with a great idea, he says he would see the product on the shelf—with someone else's patent on it. That hurt. To add insult to injury, David would then go back into his lab book and write down the date he saw his idea for sale in a store next to his idea in his lab book. That hurt even more.

David knew he had to end this torture. He decided he'd go ahead and patent his next idea—the Wrist Rest.

At the time, he was working for Microsoft and suffered from—you guessed it—carpal tunnel syndrome. Spending hours typing on the keyboard, his hands were hurting. His co-workers' hands were hurting, too, so he had a captive audience for every prototype of his Gel-eez Wrist Rests and Mousepads. It took a year of fiddling with various prototypes until David came up with

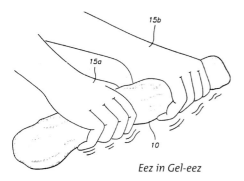
Eez in Gel-eez

a version that had the qualities he wanted and that could be manufactured. Then David hit on a "brrrreakthrough" idea. He used a gel material that could be refrigerated. That way, it could act as a cold compress while the wrists are held in a healthy position—something this kinesiology grad ought to know.

The cold compress reduces inflammation and helps relieve the pain of carpal tunnel syndrome. Cool, huh?

That was ten years ago, when David was 35 years old. After getting the patent, he licensed Gel-eez products to Case Logic. For seven years, Case Logic sold it around the world. His royalty checks totaled half a million dollars over seven years—that figures out to over $70K a year! Not a bad return for a pain in the neck—er, I mean, wrist.

He thought the Gel-eez would stay on the market longer, but the company that bought Case Logic dropped Gel-eez in order to reduce its product range. He recently got the rights back, however, and now sells it on his own website.

After his success with Gel-eez, David thought that this invention biz was pretty easy. His next invention, an ergonomic handgrip, came about due to a kitchen snafu. When David's wife, Galina, loaded a plastic-handled chopping knife into the dishwasher, it fell to the bottom close to the heating element. After going through the wash cycle, the handle was melted. Unloading the dishwasher, David saw the mangled handle. Then he held it. It fit like a glove. At that moment, David got a sharp idea.

David cut a path to the patent office and found there was nothing like it. He was off and running to develop both left-handed and right-handed versions of this grip and has since patented both of them.

David named this ergonomic handle the Golden Grip. The first thing he put the handy handle on was knives, to make it easier to grip and cut. He hopes it will prolong independence for elderly people and possibly arthritis sufferers, too. He's having biomechanical engineers at the University of South Dakota do a study on it, comparing his Golden Grips Cutlery against other handles.

These days, David is looking at various tools as the next frontier for Golden Grips from his office in Sammamish, near Seattle, Washington. Tools he may include are wrenches and handsaws. He realizes the grip won't work on all tools, such as screwdrivers. It allows a user to hold something steady and apply more force, but it doesn't allow them to twist things.

When it comes to money, Golden Grips has not had the Midas touch. It's still a work in progress, but luckily, David's first invention was profitable. He's been living off of Gel-eez, plus he's had a regular job until three years ago. While it was great to bounce ideas off his fellow inventors at Microsoft, he says he got tired of making Bill Gates a billionaire—again—and again—and again. He wants to take a stab at making himself a millionaire instead. Maybe his Golden Grips knives will make the cut.

Getting a grip

WordLock™

Letter Perfect

"I was stupefied. It seemed obvious that words are so much easier for people to remember . . ."

By substituting words for a random string of numbers, it's a safe bet that Todd has put a new spin on locks.

When Todd Basche and his wife, Rahn, moved into a house with a pool, they put a fence around it to keep their 10-year-old son, Alex, safe. The fence had three gates, so Todd bought three locks with keys. "It became clear right away that we weren't going to be carrying keys in our bathing suits."

Todd then bought combination locks. He picked a numerical combination he figured all of the adults could remember. He was wrong. Todd thought, "This would be so much easier if it used words instead of numbers. I'll go to the store and get one that uses words."

He searched all the hardware stores, certain that it must exist. Then he searched online. He couldn't find one. "I was stupefied. It seemed obvious that words are so much easier for people to remember; why not a combination lock that uses words instead of numbers?"

He decided to make one; actually, he didn't have much choice. One day,

Todd's wife came home and laid a new product on the table. It was an idea Todd had worked on years ago, then given up on. "Rahn said, 'You have good ideas and you're the most intuitive person I know. Trust your gut.' I vowed to pursue my next idea."

Good as his word, he got to work on the WordLock, then started to see why they didn't exist. "If I made it the simple way, I'd need twenty-six tumbler positions for A to Z. That would make a large lock. And it would need a mechanical mechanism with twenty-six positions, which would be cumbersome and is not even made. Everything out there uses a 10-position lock."

With over twenty-five years of experience in product development, including heading the team at Apple that developed iTunes, iPhoto, and iMovie, Todd tackled the combination from a high-tech perspective. He wrote and patented software that would maximize the number of

Turning a phrase

words the WordLock can spell using a 10-position lock. Then he got a lock on the WordLock name with a trademark.

For four years, Todd worked on his idea part-time. "I'd work at the office all day, come home and spend time with my family, then work on my invention after the family went to bed, between 10:00 p.m. and 2:00 a.m." Four years after filing the application, the patent was granted—on his son's birthday. "That's the great part of the story—that the inspiration for the product was to keep Alex safe. And out of all the days in those four years, it was issued on his birthday."

After being voted the number one product in the STAPLES® Invention Quest contest out of 10,000 entries,

the WordLock was licensed by Staples. "When Staples and I go to make a trillion padlocks, we develop a list of words that the WordLock will spell. You can have theme-based locks, such as sports or movie words. You can have locks in different languages. We baked in a bunch of French words, because we knew we would sell the lock in Canada. You can determine how many letters to use, with a space before or after the word. The locks spell thousands of words as well as all kinds of neat expressions similar to what people use on license plates and kids use on Instant Messaging."

Sold at Staples in the U.S. and Canada and on their website, Wordlock has great sales. "I can't give a number, but the Staples guys were blown away. It outsells the Master Lock two to one." In addition to the original padlock, Todd came out with several models, including a luggage lock, bike lock, and laptop lock.

People have really connected with it. "Kids play with it, twisting and turning the letters, searching for words. People play with it like a puzzle." The WordLock has been written about in *Entrepreneur* magazine, *InStyle, Reader's Digest,*

Parents, and *New York* magazine, among others. "It's had remarkable coverage from a PR standpoint.

"In the history of locks, nobody had done it. Companies that have been around a long time often get complacent. Why innovate? I give a talk on the art of everyday things. To create the WordLock, I combined clever thinking with today's technology. This process works on lots of products."

Todd is currently working on other products used around the home. He wants to manufacture his next product himself instead of license it, but is looking for retailers who would be interested in these new products.

Todd recognizes that WordLock and his other ideas may deliver apparent solutions. But they're only obvious after Todd shows them to the world. "I could have named my company 'Duh,' but it was taken."

GET YOUR GADGET GOING

So you think you have an idea for an innovative product of your own. Now what do you do?

That's a very good question, because the brainstorm is just the beginning. The road of the gadget inventor is a long one with lots of turns, potholes, and dead ends. It's not easy. It takes confidence in your idea and stubborn persistence. But, in the next few pages, you'll find some information to help get you started.

Do Your Homework

If you have a product idea, check it out on your own before you do anything else. Spend a few hours and search the Internet to see if your idea isn't already out there. Try various keywords and word groupings. Try a few different search engines and also do a search on the U.S. Patent and Trademark Office website: www.uspto.gov/ main/site-search.htm. You should really get to know your product area well and any possible competition you may face.

When it comes time to get a patent or trademark on your invention, it's wise to work with a registered patent attorney. If you don't deal with a patent attorney, you'll jeopardize your patent protection and might be on your way to getting scammed. But make sure you find a reputable attorney. Contact your local bar association, check to see that the attorney is not under any government investigation, and get at least two references.

For filing a trademark alone, most attorneys will charge under $1,000. For filing a patent, you can expect to pay between $3,000 and $5,000.

Patents 101

What is a patent? Good question. According to the U.S. Patent and Trademark Office site, "A patent for an invention is the grant of a property right to the inventor. Patents are granted for new, useful and non-obvious inventions for a period of 20 years from the filing date of a patent application, and provide the right to exclude others from exploiting the invention during that period."

There are three different types of patents available under U.S. law:

A *Utility Patent* is for the functional or structural aspects of an apparatus, composition of matter, method, or process. This means you are filing a patent on how something works.

A *Design Patent* is for the ornamental design of useful objects. Here, it is all about the aesthetic look of an object.

A *Plant Patent* is for a new variety of living plant, usually flowers. That really says it all.

Trademarks and Service Marks

Now, depending on the idea or product, a patent may not be your only protection.

Many products that don't qualify for a patent can still qualify for a trademark or a service mark.

According to the U.S. government, "A trademark includes any word, name, symbol, or device, or any combination used, or intended to be used, in commerce to identify and distinguish the goods of one manufacturer or seller from goods manufactured or sold by others, and to indicate the source of the goods." To put all that another way, a trademark is basically a brand name. And here's an interesting fact: A product doesn't even need to exist to have a trademark. Trademarks can be filed before a product is introduced into the marketplace.

A service mark, similar to a trademark, "is any word, name, symbol, device, or any combination, used, or intended to be used, in commerce, to identify and distinguish the services of one provider from the services pro-

vided by others, and to indicate the source of the services." Well-known service marks include "Federal Express℠" and "Jiffy Lube℠."

To learn more, I talked with Michael S. Neustel, an intellectual property (IP) attorney and owner of Neustel Law Offices, Ltd., in Fargo, North Dakota.

> Very often companies have both trademarks and service marks on the same mark. For example, the "Disney®" mark acts as a trademark when affixed to a pair of Mickey Mouse ears in a store, but as a service mark when used with amusement services such as Disney℠ World.

Michael is also the cofounder of the Northern Plains Inventors Congress (ndinventors.com). He told me that too many inventors focus only on getting a patent and miss the opportunity to file for trademark protection. He says, "If you've come up with the perfect name for your product, it really captures what it is and adds value to your invention, you should consider trademarking the name, even if you can't patent the product."

Making *Your* Mark

For a name to be trademarked, it must be distinctive. Usually, the more distinctive a trademark, the stronger its legal protection. Scott explains that trademark distinctiveness is described in four categories:

Arbitrary/Fanciful means that the mark has no relationship to the underlying product. For example, the words "Lexus®," "Kodak®," "Exxon®," and

"Apple®" have no direct relationship to their underlying products. Trademarks in this category have a very high degree of protection.

Suggestive Trademarks indirectly describe a characteristic of the underlying product or service. For example, the word "Coppertone®" is suggestive of suntan lotion, but this does not directly describe the product. Like arbitrary or fanciful marks, suggestive marks are distinctive and also get a high degree of protection.

Descriptive Trademarks directly describe a characteristic or quality of the underlying product. For example, "Holiday Inn®" is the trademark of a hotel company catering to vacationers. "Holiday Inn," in itself, could not be a trademark. However, since "Holiday Inn" has become recognizable from years of advertising and use, and has acquired so-called "legal distinctiveness," it is now a legally strong trademark.

Generic Terms such as "baseball glove" or "tennis racket" cannot be trademarked. You cannot trademark "Bicycle, Inc." as the name for a bicycle manufacturer. And here's a fun fact: Many common terms such as "escalator" and "aspirin" originally were trademarks that fell into the public domain. That just proves how weak the protection is in this category. That's why you'll see many brands in capital letters or with a little

"TM" after the name. It serves as legal notice to consumers and competitors of their claim to the mark, which identifies them as the exclusive supplier of a specific product or service.

I talked with Andy Gibbs, the CEO of PatentCafe.com, who says that if you are "trading" based on a brand (e.g., PatentCafe®), then your consistent use of a TM (notice of trademark use prior to Patent Office registration) or ® (only after the USPTO grants registration, hence "registered trademark") serves as commercial notice that you are the origin of such products or services.

If you fail to consistently claim rights to the mark as it may be *connected to your products or services,* then you inadvertently allow the name to become a generic description. In other words, Andy says, "Without the brand distinction, a consumer may confuse any similar product or service with yours; then the trademark gets reduced to a generic description of the product or service." And when you allow the trademark to become a generic description, you lose the right to enforce it.

Terms such as "confusingly similar" or "likelihood of confusion" are used to describe the legal boundaries of a trademark's protection (or infringement). In other words, if consumers are likely to be confused by a different trademark, or the likelihood of confusion exists, a trade-

If you came from another planet and had never seen an iPod, would you instantly know what an iPod was from just hearing the name? Apple has invested in the creation of a meaning for the word iPod, and in doing so has created a strong, recognizable brand, or trademark, for its MP3 music player.

mark may have been infringed. As an example, if a flying disk company came out with what they called the "Fizby," they would likely infringe "Frisbee," since a buyer in a store would assume that the "Fizby" is a "Frisbee."

This is why trademark owners must be diligent about making sure their trademark does not slip into generic terminology. For example, Xerox has, over the past ten years, worked very hard to "reclaim" its brand name after witnessing widespread generic use of the name—"Can you make me a Xerox copy of this letter?" Now Xerox polices the use and insists that you say, "Can you make me a XEROX BRAND copy of this letter?" They don't want Xerox to meet the same fate as "Aspirin."

Velcro is facing the same challenge.

Competitors must call their products "hook and loop"; Velcro polices use of the brand to prevent it from going generic.

A trademark is an *extremely* powerful form of intellectual property protection because of the investment required to build the brand. Just ask McDonalds if you can borrow their golden arches, or Nike their "swoosh." They won't let you "Just Do It." And Burger King won't let you "Have It Your Way" either.

What does this mean for garage inventors? If, let's say, you've come up with a new toilet bowl cleaner, don't call it "tidy toilet." Sure, it sounds cute, but it's a very weak trademark simply because it has meaning—it is *descriptive* of the product.

On the other hand (perhaps counter-intuitively), a name such as "minute wizard" could be a strong trademark for your toilet cleaner precisely because it tells you nothing about the product. However, through advertising, you can create the association of your toilet-cleaning product with the fact that you can clean the toilet in less than a minute. Once you establish the connection between "fast and easy" and your toilet cleaner, you've created a real, solid trademark.

"Inventors Beware"

Inventors' passion for their products can cloud their judgment, causing them to drop their guard and become easy prey to hustlers. In gathering stories for this book, I heard over and over again how unscrupulous companies feed on the hopes and dreams of would-be inventors. They send out mailers and advertise on late-night television. They often offer all-in-one services, including patent filing, manufacturing, and marketing for an inventor's product. They profit by trapping unsuspecting inventors in their web of promises and false hope. Eventually, the inventors learn the truth—that the work these companies actually do falls far short of the promise and that they have little or no legal recourse. It ends up being an expensive lesson, often draining $15,000 or more from hard-earned bank or retirement accounts.

If you learn nothing else from this book, I hope you learn to beware. You've heard it before: "If it sounds too good to be true, it probably is." Companies that promise too much need to be carefully evaluated. American Inventors Protection Act of 1999 (AIPA) requires that companies inform inventors of their record of customer complaints as well as their history of success. It might surprise you that many of these invention promotion companies have success rates of less than one percent. Ninety-nine percent of all inventors who pay money to them walk away with nothing but an empty wallet.

The Business End

Just because an individual has the ingenuity to come up with a brilliant invention, there's no guarantee that same person has a brain for business. As you've read through this book, you've come across many tortured tales of good inventions gone bad. Good ideas that lack market research and a solid marketing plan often remain just that—ideas. Turning an idea into a successful product takes much more than merely figuring out how you are going to manufacture it. You need a plan to let the world know that your product exists! Inventors need to do their homework. Visit a local business school and talk to

America's Invention Protection Act calls for the U.S. Patent and Trademark Office to publish complaints that inventors have lodged against unscrupulous invention-promotion companies. Check out uspto.gov/web/offices/com/iip/complaints.htm to see if a company you are dealing with is already listed. If it is, run, don't walk, away.

some of the professors to pick up marketing savvy and specific suggestions (such as referrals to local ad agencies), or call the editor of the business section of your local newspaper. It makes sense to make a few appointments with ad agencies and public relations companies in your area—but don't just sign on after that first meeting. Listen to their proposals, then shop around. Compare and evaluate agencies before you make any commitment. Some public relations agencies, such as HWH (hwhpr.com), offer affordable online PR services for start-up companies. Investigate your options.

Where to Get Help

The road of the inventor may be long, but it is well traveled. Many inventors have been down that same path. Inventors shouldn't feel alone. There's a great deal of very useful information out there in libraries and on the Internet. A great place to start is at www.uspto.gov/smallbusiness/. If you click on patents, you'll find that the U.S. government has put a great deal of energy into making a user-friendly, information-filled site for inventors and small business owners.

Inventors should also turn to other inventors. There are more than 100 associations and organizations throughout the U.S. where inventors can meet and swap information. If there's not an inventor organization near you, consider traveling to one of the annual inventor trade shows. There, you'll be able to attend sessions where seasoned inventors will teach you how to build prototypes, how to work with your patent attorney, and how to license your invention to large companies.

Check out these groups:
- United Inventors Association USA: uiausa.com or uiausa.org
- Yankee Invention Expo: yankeeinventionexpo.org
- The Inventors Network: inventnet.com
- Minnesota Inventors Congress (resource center and annual convention): inventhelper.org
- The International Federation of Inventors' Associations: invention-ifia.ch

You can also find a listing of other groups at the National Inventor Fraud Center website: inventorfraud.com/inventorgroups.htm

Finally, don't forget about your good old Uncle Sam. The Small Business Administration has Small Business Development Centers, called SBDCs, in all fifty states. The SBDCs can offer you assistance in writing a business plan, evaluating invention promotion organizations, applying for research grants, licensing your invention, and much more. The best part is that advice from an SBDC is either free or very, very affordable. Check out sba.gov/sbdc to find out more.

OK, now you have the information, so go out and invent something. Make the world a better place. We live in a society that hungers for new things, so feed that hunger—come up with the next must-have gadget. You may not make millions, but you just might make history.

INDEX

Photo Credits

Jack Duestch: page vii, 2–3, 8–9, 13, 16–17, 18–19, 20–21, 24–25, 29, 30-31, 32, 41, 44–45, 46–47, 52–53, 54–55, 56–57, 59, 60–61, 62–63, 64–65, 66–67, 72–73, 79, 82—84, 85–87, 94–96, 97–99, 104–105, 110–111, 116–117, 119, 121, 125, 130–131, 134–135, 136–137, 138–139, 146–147, 148–149, 153, 162–163, 166–167, 168–169, 172–173, 176–177, 184–185, 187, 190, 195, 198, 200–201, 206–207, 208–209, 212–213, 218–219, 220–221, 222–223, 224–225, 231, 232–233, 237

istock .com photo: page 26, 38–39, 51–52, 60, 66, 76, 77, 98–99, 128, 145, 151, 152, 186, 220

Inventors, their families and friends: photographic and other illustrative materials supplied for use, with permission, by the individual inventors.